W. J. Moore

Health in the Tropics

Or Sanitary Art Applied to Europeans in India

W. J. Moore

Health in the Tropics
Or Sanitary Art Applied to Europeans in India

ISBN/EAN: 9783337061159

Printed in Europe, USA, Canada, Australia, Japan

Cover: Foto ©berggeist007 / pixelio.de

More available books at **www.hansebooks.com**

HEALTH IN THE TROPICS;

OR,

SANITARY ART

APPLIED TO

EUROPEANS IN INDIA.

BY

W. J. MOORE,

LICENTIATE OF THE ROYAL COLLEGE OF PHYSICIANS, EDINBURGH ; MEMBER OF THE ROYAL COLLEGE
OF SURGEONS, ENGLAND ; BOMBAY MEDICAL SERVICE ;
FORMERLY THREE YEARS RESIDENT-SURGEON AT THE QUEEN'S HOSPITAL, BIRMINGHAM.

LONDON:
JOHN CHURCHILL, NEW BURLINGTON STREET.
MDCCCLXII.

" The largest material resources involve the heaviest moral responsibilities, and service rendered to the humblest and meanest of mankind is rendered to the Author of our common being."—LORD STANLEY'S SPEECH ON INDUSTRIAL RAGGED SCHOOLS, *at Liverpool, Jan. 7th*, 1862.

J. E. ADLARD, PRINTER, BARTHOLOMEW CLOSE.

PREFACE.

No less authority than Ranald Martin has advanced the decided opinion, that a Sanitary Manual, containing in a clear and concise form the great truths and principles of hygiene and sanitary art, should be published "by authority," and become the guide of both military and medical officers.

In conversation with many gentlemen of both classes, the absence of such a work has been allowed to be an existing want.

A volume on some such plan as the one now deferentially offered (perhaps, however, requiring some deductions, and certainly wanting additions), appears calculated to obviate the deficiency. And to make the work complete, the obsolete and standing orders relative to each subject, together with the reasons for the abolishment of the former and the publication of the latter, should follow their appropriate chapters, or otherwise be placed in an appendix.

I at first intended to have added these orders and references, but at an up country station there are not facilities for such a compilation.

Time, and the sanction of authority to search records, would be required before all such regulations could be searched out and methodised.

But it is not only to a particular class that the present volume may be expected to prove useful. All ranks of Europeans in tropical climates are exposed, *cæteris paribus,* to the causes of disease which produce the great mortality amongst the European soldiery; and, therefore, the same care and attention to sanitary measures are required by the former as by the latter.

The application of the sanitary regulations laid down in this work may be easily directed from the soldier to the civilian, in any condition of life, by the unaided perception of the latter.

W. J. M.

Note.—Owing to the circumstance that this book, written in India, is printed in London, and the consequent absence of the author's revision, a few typographical errors can scarcely be avoided. As was stated, however, in one of the reviews of my former publication, entitled, 'A Manual of the Dieases of India,' it is hoped that in the present instance also, "such errors," should they occur, " will not diminish the general utility of the work."

CONTENTS.

CHAPTER XII.

ON CLEARING.

CHAPTER XIII.

ON BARRACKS.

CHAPTER XIV.

ON HOSPITALS.

CHAPTER XV.

ON CONSERVANCY.

CHAPTER XVI.

ON WATER.

CHAPTER XVII.

ON CHOLERA.

CHAPTER XVIII.

ON THE SOLDIERS' DIET.

CHAPTER XIX.

ON INTEMPERANCE.

CHAPTER XX.

ON SCURVY.

CHAPTER XXI.

ON PROPHYLACTIC MEDICINE.

CHAPTER XXII.

ON SYPHILIS.

CHAPTER XXIII.

ON DRESS.

CHAPTER XXIV.

ON THE EMPLOYMENT OF THE SPARE HOURS OF THE EUROPEAN SOLDIER IN INDIA.

CHAPTER XXV.

ON CAMPS AND MARCHING.

CHAPTER XXVI.

ON THE DESPATCH OF TROOPS BY SEA.

HEALTH IN THE TROPICS.

CHAPTER I.

FEW of the great scientific advancements from which we now reap manifold advantages have been established without encountering much opposition, oftentimes virulent, contemptuous, and sarcastic. In whatever department of science or art we take a retrospective view, we find that its history most frequently develops one continued struggle against the prejudices and opinions of the age. In proof of this may be instanced the establishment of the spinning-jenny by Hargreaves, of the capabilities of the locomotive by Robert Stephenson, of the circulation of the blood by Harvey, and of vaccination by Jenner.

1

It has been observed, that the strongest evidence of human progress is the conquest of science over error and superstition; and during recent periods, the more general diffusion of liberal education has allowed the demonstration of many startling projects in art and science, which, "sixty years since," would have been utterly scouted as impossible. Hence it has more than once happened that new designs and discoveries have been readily, indeed enthusiastically, received. Of these, the electric telegraph and chloroform may be cited as instances.

Although attention to sanitary regulations has proved productive of wonderful results, and notwithstanding these results are demonstrated to be certain consequences of such care, still it is unhappily not the less a fact that sanitation and hygiene are *not* among those projects which have been readily received by mankind. Those essential arts are still struggling to arrive at maturity.

And why is this? Chiefly because sanitary knowledge simply offers health, and not immediate wealth or honour, as rewards to her disciples; secondly, because ignorance frequently causes her teachings to be regarded with derision or contempt; and thirdly, from the fact that sarcasm has been directed towards sanitary improvements even by some among the ranks of the medical profession!

In honour, however, to the profession, let it be known that those who, by their writings, have tended to delay the general progress of sanitary reform are, with few exceptions, to be sought for among the reviewers; in other words, among those who write anonymously, but whose efficacy for good or evil—owing mainly to the use of the authoritative WE—is, perhaps, greater than the influence of the independent author.

Thus, in the 'Edinburgh Medical Journal,' November, 1860, in the review on the 'Medical Regulations for the Duties of Medical Officers of the Army,' published 1859, the reviewer writes :—" In connection with the great talk that we have heard about sanitary science, it is interesting to note what that is, when put in the form of precise regulations. It is found to consist in opening windows, in keeping privies clean, in enforcing a circulation of air about beds and bedding, and in attending to drainage, diet, and water supply —all matters of great importance, of transcendant importance, much more important than physicking after the manner of English apothecaries; and, as the world is beginning to find out, involving considerations which are revolutionising the whole practice of medicine, but hardly to be put forward as a science. Sanitary science has been one of the current cants of the last few years; statistics has been another."

Here we have the reviewer, after giving a most meagre list of a few of the least important matters which sanitary knowledge embraces, objecting to its being classed as a science; and yet in the same paragraph stating that this to which he denies a name is revolutionising the whole practice of medicine; yet medicine, in her present state, even the reviewer would scarcely designate in less honorable terms than science.

As regards statistics, in India at least, let the researches of Waring, of Coles, of Ewart, and others—the statistics of disease in the Indian armies, in the Indian gaols, speak for themselves !

The arts of hygiene and sanitation are not based on any special science of their own, but are the application to practical ends of principles derived from other sciences. To become a teacher of either one or the other—to apply such principles either to individuals or communities—demands, as Lord Herbert observed, a considerable portion of that special knowledge which appertains to the " physician, physiologist, geologist, meteorologist, topographist, chemist, engineer, and mechanic."

Whether henceforth sanitary progress will be classed as a

science or an art, the practical importance of the principles
involved remains the same, and their neglect, as heretofore,
will be followed by the loss or diminution of that greatest
earthly blessing, health—

" Which is than kingdoms far more precious."

Under such circumstances, the sanitary reformer may well
decline a contest for a name, and be content to call his progress " art."

The almost entire extermination of scurvy, dysentery, and
intermittent fever in London, for instance, is one of the most
striking facts in sanitary progress.   One hundred years ago,
agues and malarious fevers were among the most common
diseases, not only of the metropolis, but of the whole of England.   At the present time, owing to the almost universal
drainage, exact cultivation, and more general adoption of
sanitation, these diseases have, excepting in a few isolated
spots, entirely vanished.   In the 17th century, the annual
deaths from dysentery and bowel complaint fluctuated in
London from one to two thousand; now the annual average
does not exceed from twenty to one hundred! And it cannot
be doubted that we may confidently look forward to an equally
complete extermination of the whole class of zymotic disease.
Indeed, typhoid fever has come to be the accepted test of the
sanitary condition of a locality, and a diminishing mortality
from such maladies a certain sequel of attention to the principles which govern the action of sanitary regulations.   So
close is the connection between defects of sanitation and the
disease just named, that Dr. Murchison has, with the assent
of Dr. W. Farr, denominated it pythogenic fever, or fever
generated by dirt.[1]

Passing from civil to military matters, we again find the
strongest evidence of the paramount importance of sanitation,
and of the truth of the great principles of hygiene, in the

[1] 'Lancet,' Dec., 1861.

fact that the sanitary regulations which have been introduced, and are yet in progress, have raised the health-rate of the soldier in temperate climes *from that of the baker and compositor to the standard of the middle-class civilian!*

These sanitary arrangements were *begun* at home; but, Dr. Balfour's account, published in the late 'Army Statistical, Sanitary, and Medical Report,' demonstrates that in Canada, the North American stations, in the Mediterranean, in the West Indies, in St. Helena, in the Mauritius, and in Ceylon, improvements are observable. India, where we have some 80,000 European troops is *not* included in the report; *it still remains to be dealt with.*

What was it, excepting the general absence of sanitary regulations, which, in the earlier period of the Crimean war, caused the decimation of an army, and the lamentations of a nation? And what but sanitary regulations enabled us to present to the admiration of the whole civilised world, during the latter portion of the same campaign, an army whose health, and therefore efficiency, was contrasted favorably with legions employed on garrison duty at home!

Nay, what was it but the thorough adoption of sanitary regulations which enabled us in the late China campaign to afford what has been designated " the almost unparalleled spectacle" of an army moving rapidly in an enemy's country, and that country the sea-coast of a semi-tropical district, in the highest condition of health and efficiency, " where, thanks to the excellent arrangements of Dr. Gibson, the admirable hospital ship of Dr. Mapleton, and the efficiency of the medical officers, the soldier's health was equal to the highest average of the civilian at home, and the mortality just one tenth less than the mortality of the army in the Crimea, during the first seven months of that disastrous campaign?"

Regarding the China war, General Peel[1] states :—" Let

[1] 'Home News,' Oct. 18th, 1861.

me observe, that there never was any war which reflected greater credit on every one connected with its management. The organization, the equipment, the arrangements made for the comfort of the troops were really excellent."

As Mons. Thiers remarks, in ordinary histories we see only armies completely formed and ready to enter into action—the effort it has cost to bring the soldier to his post, to train, to feed, and cure him, if sick or wounded, are lost sight of. The nation sympathises deeply with the result of a battle; the slaughter of a few hundreds by sword or ball excites the feelings to the utmost; but the tenfold loss occasioned by unnecessary exposure to the burning sun; by the unhealthy bivouac; by the malarious encampment; by the badly chosen cantonment; by the ill-adapted barrack, is looked on as the necessary consequence of military service or warfare. Let us hope that this has now ceased to be applicable to the British nation.

The lessons of experience are often dearly bought, and quickly forgotten; but, if a lesson were ever purchased by a nation at a cost of life and treasure which should redeem it from oblivion, it is, that the means of preserving the health of the soldier in the field and in garrison are precious beyond thousands sterling. The loss of many hundreds of soldiers, each of whom has cost the state £100 to train to his duty, has resulted from the neglect of sanitary regulations, and, it may be added, a long depreciation of the medical department. All these facts are now matters of public notoriety, and, in future, it may well be demanded, why any military body suffers from a large per-centage of disease; and with not less force because it is becoming generally understood that disease, especially zymotic disease, *is preventible, or at least mitigable.*

In the speech above referred to, General Peel proceeds to remark:—" I have alluded to the sanitary arrangements made for the comfort of the troops in China, and I trust I may be permitted to avail myself of this opportunity of expressing my regret for the loss we have sustained in the death of Lord Herbert. No man was more anxious than he to promote the welfare of the British soldier; and his great object was to impress on Parliament that prevention is better than cure. It was that conviction which induced Lord Herbert to establish a medical school with a hospital attached, in which young men might be qualified to become members of that body, than whom there is no more worthy class of men—I mean the medical officers of the army."

To Lord Herbert we are also indebted for the ' Code of Regulations for the Medical Officers of the Army,' published in 1859, which, if not perfect, is an immense advance in the required path.

Paragraph 28, page 81, of these ' Regulations,' constitutes the senior medical officer of a station " *ex officio* sanitary officer."

Prior to the promulgation of this order, sanitation can scarcely be said to have had a recognised existence in India; and even now *representation* is all the power allowed to the medical officer: *he may forewarn, but, like Cassandra, his prophetic voice is frequently not heeded until too late.*

As Lord Herbert, however, observes, " The constitution of an army requires that the commanding officer should be supreme, as he is responsible within his command. It would be contrary to every principle of discipline that any other officer should dictate to him what he ought to do. If the education of military officers comprehended a knowledge of the principles of sanitary science, commanding officers might safely be left to their own judgment in adopting sanitary precautions for protecting the health of the men. Such, however, is not the case; and the problem with which we have to deal is, how to supply the commanding officer with competent advice on which to form his judgment, and yet keep his supremacy absolute and *intact.*"

This has been attempted by requiring that the medical officer should state to the commanding officer in writing whatever representation he has to make on any matter affecting the health of the troops, and that the commanding officer should take such recommendation into consideration, and act upon it or not, as he thinks right.   But in the latter case, the commanding officer should state his reason for non-compliance in writing, so as to insure that the advice shall not have been inconsiderately rejected when the matter comes under review by superior military authority.

In my work entitled ' A Manual of the Diseases of India,' page 37, I wrote—"Would that natural philosophy and sanitary science entered more largely into the studies of the military officer." And *it is only by such education of military officers that the full advantage from sanitary science and hygiene will be established in the army.*   In the meantime high *special* sanitary might be more immediately combined with superior military authority in reviewing the rejection of recommendations made by subordinate officers.

" 'They do these things better in France," is a remark too frequently most undeservedly applied, but it would appear to be apropos to the subject under consideration.   Regular lectures on hygiene and sanitary matters have been lately delivered at the St. Cyr School, in France, where young gentlemen are trained for commissions in the army.   As the editor of the ' Dublin Medical Press' observes, "This innovation will have very beneficial results in initiating military officers into the principles of hygiene, which may by them be applied for the advantage of their men."[1]

That Lord Herbert appreciated the difficult position of medical officers, is fully evident from the following : his lordship wrote :—"When a medical officer goes to the general commanding, who, under a tropical sun, up a river surrounded with swamps, is feeding his troops with salt pork, and tells

[1] 'Dublin Medical Press,' Dec. 26th, 1861.

him that unless he gives them fresh meat and vegetables they will be down with scurvy and fever, he does no more than his duty, and what it is imperative that he should do. But if he is met by the man in authority by the rejoinder— 'Sir, when your advice is wanted it will be asked for,' he probably vows never again to expose himself to such a rebuke. Six weeks after he is called upon to cure disease which is not curable at all, or not curable in time, though care and prevention a few weeks earlier might have obviated much of it."

Hence, as it is and ever will be contrary to every principle of discipline that a medical officer should dictate what the commanding officer is to do, the principles of sanitary science should form part of the educational course of every military officer. Having been made aware of the truth of those principles, there would be little danger of any superior officer questioning their practical application in matters of detail, the due observance of which must always be initiated by the professional knowledge of the medical officer.

The remarks of the Hon. and Rev. W. H. Lyttleton[1] "On the Practical Teaching of Sanitary Science in National Schools," may be applied to the military educational establishments:— " So long as the poor are unwilling to obey sanitary laws, we shall be but, Sisyphus-like, constantly rolling stones to the top of the hill, which will immediately roll down again. Let, then, the principles of sanitary science, the practical consequences of good and evil, which flow from obeying or neglecting its laws, be thoroughly taught in *all* schools."

If this were the case, the extreme absurdity of appointing A BOARD CONSISTING OF ONE lieutenant-colonel and two barristers to draw up the ' Report on the Origin of Yellow Fever in Bermuda in 1853 ' would not be so self-evident. As it was, it appears an admirable example of placing the wrong men in the wrong place.

[1] ' Sanitary Review,' June, 1855.

If ever a committee or board of the kind is to hold its true position, it must consist of men under competent presidency, who are acquainted with the principles of sanitary administration. Its members, however, must not have merely a vague idea of the causes of disease, but some at least among them must know, too, *the nature of disease itself*, and we shall not receive the full benefit from sanitary knowledge until these broad and common-sense views are acted upon. Granted that the Bermuda board were *au fait* as to sanitary principles, it can scarcely be expected that even one among the number was conversant with the nature of disease. *Nec scire fas est omnia—*

> "One science only can one genius fit,
> So vast is art, so narrow human wit."

Although sanitary art is only now being brought so prominently forward, and notwithstanding it has but latterly been made a recognised part of our military system, individuals have not been wanting, who, even in remote ages, acknowledged and taught the great principles of hygiene.

Without reviewing the sanitary teachings of Moses, Lycurgus, and Hippocrates, followed by Galen, Oribasus, Ætius, Paulus Ægineta, King Alfred, and, in later days, Lord Bacon, Sanctorius, Ramazzini, Frank, and others, it will be sufficient to state, that more than one hundred years ago there were military surgeons who understood, and indeed by their writings taught, many of the great principles of sanitation.

Thus Pringle[1] knew and acted upon such knowledge, that the best and cheapest of all medicine is a boundless profusion of fresh air, and would "rather shelter and treat sick soldiers behind a hedge-row, than in the reeking wards of a military general hospital."

"Air," says Pringle, "that is become putrid by confine-

[1] 'Pringle's Campaigns.'

ment, stagnation, and animal effluvia, is the cause of sickness the most fatal and least understood. These destructive steams work like a ferment, and ripen all distempers into putrescence and malignity."

Again, as regards overcrowding in public buildings, Sir Gilbert Blane, who was the first to compare the treatment of the sick in hospitals with the treatment in private practice, records the general law, observable both among the lower animals and men, that when large numbers are congregated in ill-contrived tenements, so that the exhalation and excretion from the living body are not completely removed, then disease is produced. Thus glanders arises among horses; distemper among dogs; and typhus in gaols, hospitals, or ships, among human beings.

Both past records and recent experience show that, when a ship, or gaol, or barrack, or an hospital, is overcrowded, deadly fever, or hospital gangrene, erysipelas, tetanus, or the so-called constitutional, irritative, or surgical fever will arise.

If to overcrowding and bad air, or want of ventilation, be added uncleanliness, deficient clothing, or improper food, with so much the greater facility will these diseases be generated.

Of the effects of personal uncleanliness or of deficient clothing in hindering and retarding the action of that important organ, the skin, it cannot be necessary to dilate; and, as Dr. Ewart[1] observes, "There is scarcely a disease to which the human frame is liable that is not seriously influenced in its origin, development, climax, and termination, by the nature, quantity, and quality of the diet which may have been employed."

Thus the healthiness of most prisons in the United Kingdom has of recent years been very greatly increased by the sanitary improvements which have been introduced upon the recommendation of medical officers; and the same may

[1] 'Statistics of Indian Gaols,' p. 61.

be said of many hospitals, workhouses, &c., from which typhus fever has been abolished.

Thus it was set forth in the memorial of physicians and surgeons to Lord Panmure, dated April, 1855, that the health of the inmates of such establishments depends mainly upon the observance of certain hygienic laws which the science of medicine has pointed out and experience has proved to be necessary for the due performance of the functions of life.

But it would be a vain attempt to preserve human health by confining attention to the *interior* only of a building; hence sanitary art comprises also the condition of the external atmosphere, and all the numerous accessories which exert an influence on that atmosphere. Thus the different essentials which unite in producing any given climate, the direction of the winds, the presence or absence of ozone, the rainfall, the temperature, the latitude and longitude, the presence or absence of rivers, marshes, forests, jungles, mountains, the geological formation of the ground, cultivation, the situation of trees, structure, position, and condition of buildings, drainage, and, in short, both the labours of man and the works of nature, *all* require due consideration, under the head of sanitary science.

When to all these are added *special sanitation* required for the soldier, under the varied and widely different positions in which he is placed in the field, on the march, in the transport, and in the tropics, it may be justly stated, that there is no branch of science requiring a more extended knowledge than sanitation and hygiene.

General Peel states —"It is not from war that the soldier suffers most. It is from the constant change of climate, and from the diseases to which the quarters in which he is from time to time stationed expose him."

Marshall Saxe, a high authority in such things, was in the

1 'Home News,' Oct. 18th, 1861.

habit of saying that, to kill a man in battle, the man's weight in lead must be expended. The 'Lyons Medical and Surgical Gazette' states that this ·fact was verified at Solferino, even after great improvements in fire-arms. The Austrians fired 8,400,000 rounds. The loss of the French and Italians was 2,000 killed, and 10,000 wounded. Each man hit cost 720 rounds, and every man killed cost 4,200 rounds. The mean weight of a ball is one ounce. Thus we find that it required, on an average, 272 rounds to kill a man.

Colonel Hodgson[1] informs us that the British soldier who now serves in Bengal one year encounters as much risk of life as in three such battles as Waterloo.

Such being the case, such being the danger from ⌊battle, and so much greater mortality arising from climate, it may be truthfully stated, with regard to the application of sanitation to the soldier in tropical climates, especially in India, that the economical possession of the latter country, rests mainly on sanitary art; for " what is the sword without the strong arm." With reference to the adoption of sanitation to the European army, now so necessary in the East, we may well exclaim—

" In te omnis domus inclinata recumbit."

The manner in which we must consent to hold this country is beginning to be more clearly seen. It is perceived that there must be no expectation of brilliant individual fortunes, and that the country must become more tolerable and more habitable for those who venture to it. It is also beginning to be appreciated that health in India is to be promoted by the very same means and by attending to the very same considerations which are admitted to be those which regulate sanitary matters in other countries, bearing always in mind climatic peculiarities.

In a lecture on military hygiene, delivered at the United

---

[1] Col. Hodgson's ' Military Miscellany.'

Service Institution at Poona, September, 1861, Dr. Fraser
stated that the Sepoy army of India is the healthiest in the
world, and the only military force wherein the mortality is
not greater than that of the population from which it is
drawn.   This is ascribed to the separate huts in which the
men live, and to their removal from dirty and crowded
village habitations into spacious cantonments, and, in comparison with their former mode of life, under efficient sanitary
control.

Deducting a certain mortality which must ever occur to
the European residing in a tropical climate, that which has
resulted in the native army may be as satisfactorily obtained
in the British legions.   In the meantime, until sanitation
rules with imperial sway, the words of the poet must remain
applicable to Europeans in India :

> " Art is long and time is fleeting,
>   And our hearts, tho' stout and brave,
> Still like muffled drums are beating
>   Funeral marches to the grave."

What, therefore, can be more honourable employment for
a medical officer in the service of the State, than a constant
endeavour to improve the sanitary condition of the British
soldier ?   And what more than such endeavour will enable
him to say, at the termination of his somewhat hard career,
" I have done my duty."

Moreover, where shall be found a body of men more deserving
of care than those composing the rank and file of our army ?
—" the race who (Miss Nightingale writes), with that handful
of men, defended their trenches at Sebastopol as the Greeks
held their position at Thermopylæ, and who, when dying of
slow torture in hospital, drew their blankets over their heads
and died without a word, like the heroes of old.   I have seen
men dying of dysentery, but scorning to report themselves
sick, lest they should throw more labour on their comrades,
go down to the trenches and make the trenches their death-

bed. I have known intimately the Sardinian soldier, the French soldier, and the British soldier. The Sardinian soldier was much better appointed than we were; the French were both more numerous and much better appointed, and more accustomed to war; yet I have no hesitation in saying that we had the better military spirit, the true volunteer spirit, to endure hardship for our country's sake."

And that this is not an exceptional case, let our Indian battle-fields bare witness, from Assaye to Delhi!

CHAPTER II.

ON THE CAUSES OF ZYMOTIC DISEASE.

Definition of Contagion and infection—Origin of Disease—Soiled Air—Composition of Air—Respiration—Effects of Soiled Air—Meteorological Conditions—Ozone—Exact Causes of particular Diseases not known—Kindred Properties of Disease—Contagion—Infection—The Windsor Epidemic—Extension of Miasm—Value of a Respirator—Conclusion.

FROM the foregoing remarks it will be evident that sanitary art comprises, as one of the principal items, a knowledge of the origin and dissemination of zymotic disease, and, therefore, of the vexed questions of contagion and infection.

By contagion should be understood a specific virus originating within the body.

By infection, a pestiferous atmosphere, originating without the body.

Therefore some diseases, as eruptive fevers, are both contagious and infectious.

It is commonly thought that a knowledge of such matters, concerns only the physician; and, as Dr. Headland observes, "it is far too common a habit among men, to delegate to the medical practitioner the sole and undivided control over matters of such momentous importance" as relate to the science of health. If, however, diseases really are preventable, such knowledge is of vital consequence to every member of the community.

Even supposing we believe, as was formerly so generally taught, that diseases such as smallpox or typhus fever are propagated by infection or contagion, there must still remain to be explained how the disease originated in the first instance?

If, however, we take another and more common view of the question; if we recollect that the locality of such disease is where filth, overcrowding, and defective drainage are to be found, and if we consider that *air can be soiled as well as water*, we shall have little difficulty in arriving at the conclusion that disease may be conveyed into the human system, and generated *de novo*.

The bulk of the atmosphere consists, as every one knows, of a mixture by weight of twenty-three parts oxygen, to seventy-seven of nitrogen containing about 1000 part of carbonic acid, a variable amount of watery vapour, and a very changeable quantity of the body called ozone, excepting, perhaps, in the neighbourhood of tropical rivers, where vast quantities of organic matter, the débris of a luxuriant vegetation, are rapidly undergoing decomposition; and in towns where a larger quantity of carbonic acid and a minute proportion of sulphurous and sulphuric acid, accruing from the oxidation of the sulphur in the burning coals, have been demonstrated. This identity of composition of the atmosphere is shown by the repeated experiments of chemists to be a fixed law. Whether taken from the summit of mountains or the centre of cities, chemistry refuses to recognise a difference in the composition of the atmosphere. Like water, its essential constituents are the same; but, like the latter fluid, it may contain in suspension, or, perhaps, even in solution, immense volumes of most varied and most deleterious agents.

It is not difficult to prove that animal and vegetable matter, during a state of putrefaction, disengages from its surface portions of its substance of sufficient tenuity to be invisible and to remain suspended in the air. Without referring to offensive smells, which must be material, we have other sufficiently satisfactory proofs. If a bell-glass be inverted over decomposing matter, the inner surface of the glass becomes covered with microscopical filamentous fungi. Similarly it is found that the moisture which is deposited on the roof of sewers is rich in organic matter, which must have been derived from the air of the receptacle. Again, milk will turn sour, if exposed to putrid exhalations; butchers cannot successfully dress meat near a stinking gully or dirty ammoniacal stable; flesh will not keep when the air is loaded with moisture, cloudy, and, therefore, more retentive of organic impurities.

From enquiries made under the sanction of the Board of Health, in 1854 ('Report on Cholera,' p. 127), it was found that on certain air being passed through distilled water, the appearance of hyphaceous fungi in such fluid was an invariable result. It was also found, that on passing the same air through sulphuric acid, the acid soon became dark-colored in consequence of the charring of the organic matter contained in the atmosphere.

Still more latterly, Dr. A. Smith[1], has performed some experiments by passing air through a dilute solution of permanganate of potash (the strength of which had been previously determined by its power of decomposing oxalic acid) and found that the quantity of oxidizable matter in the atmosphere was so much greater in towns than in the country, that the same quantity of the solution of permanganate which was decolorised by one bottle of air, obtained in a close court in Manchester, required twenty-two bottles to decolorise it on the hills in the neighbourhood!

In the late 'Report of the 'Lancet'[2] Commission of Inquiry into the influence of Railway Travelling upon Public Health,' some further experiments performed by Dr. Smith with permanganate of potash are also recorded. From these it appears that on the high grounds north of Manchester,

[1] Smith, 'On the Air of Towns.'
[2] 'The Lancet,' Jan. 11th, 1862.

209,000 cubic inches of air were necessary for the decomposition of a given amount of the solution; while in a bedroom in Manchester, 64,00; after the same room had been slept in, 56,000; and when the strong smell of a sewer entered the room, 8000 only were required! The latter, be it observed, is the same result which occurred when the air of a closely packed railway-carriage was examined.

Those who are conversant with the physiology of respiration, and understand the process of decarbonization of the blood, which takes place in the vesicular structure of the lungs—who know that the vital fluid is there exposed to the respired air—are well aware of the danger of a sudden or constant absorption of impure atmosphere. They who are ignorant of the anatomy, physiology, and chemistry of respiration, can only deduct from analogy; but all are more or less aware of the effect which follows the inhalation of chloroform; of carbonic acid gas from burning charcoal; of the aroma of wine in large vaults; and of the deadly results from concentrated sewer gas.

As a diluted strength of the agents above mentioned will induce their characteristic effects in a lesser degree of intensity, so will an atmosphere but partially charged with mephitic matter insidiously destroy the constitution, depress and impair the functions of the whole body, particularly of those organs with which it is brought into contact, manifesting itself constitutionally by fever accompanied by prostration, locally by affections of the respiratory and digestive mucous membranes.

Dr. Letheby states ('Report to Commissioner of Sewers,' 1858):—"One breath of the undiluted sewer gas will destroy life immediately. Mixed with common air, it will cause asphyxia and narcotism; smaller quantities produce nausea, delirium, and gradual insensibility; while further dilution induces general prostration of vital power, failure of appetite, diarrhœa of a chronic character, exhaustion, and low fever."

And these or similar results will accrue from the respiration of impure air, whether such tainted atmosphere arises from sewer gas, neglected conservancy, animal and vegetable decomposition, overcrowding in habitations, or diseased animal life.

The origin of, or rather the preparation of the system for zymotic disease having thus been demonstrated, the questions naturally present themselves, how it happens that such diseases prevail epidemically, or in greater intensity at some periods than at others, all originating circumstances apparently being unaltered.

It has been already stated that there are certain days on which meat killed quickly turns putrid, and that such seasons have been noticed as marked by humidity, closeness, or stillness of the atmosphere. It is evident, therefore, that atmospherical conditions have much to do with the activity or otherwise of the germs of zymotic disease.

Of late years, the uncertain and somewhat obscure science of meteorology—which Herschell compared to a "romance of interesting episodes"—is illuminated by unexpected lights. Science no longer remains mute respecting the approaching terrors of fearful tempests; and the globe is shown to be an immense laboratory, wherein are effected powerful combinations, which prepare and accomplish the grand perturbations of the atmosphere.

In the present state of our knowledge, there is *more* reason for referring "the epidemic constitution of the air" to the presence or absence of the matter known as ozone than to any other named agent.

That ozone, either in excess or deficit in the atmosphere, has an influence over the human constitution is proved by several recorded observations, some of which date so far back as 1845.

M. Houzeau[1] has proved that ozone reaction is destroyed

---

[1] 'Variab. Norm. des Prop. de l'Air.'

by urban contamination. Scoutellen, of Metz, found that in a close, narrow street, with a cesspool at its head, ozone was only four times developed during a period of six months, while at the neighbouring military hospital it was always present. Dr. Bœckel noticed that malaria always occurs when the ozonoscope marks zero, or the lowest possible degree, and that marsh fevers rage most severely under exactly the same circumstances. Ozone, moreover, as is well known, is rapidly absorbed by a great number of animal and vegetable substances, such as albumen, caseine, and blood, and, from its oxidizing properties, is the most powerful disinfecting agent known.

Dr. Pickford[1] remarks : " In confined places, where ozone cannot penetrate, plants and men become blanched, the skin grows pallid, the blood loses colour, lymph predominates, all the tissues soften, and serious disease of the adynamic type breaks forth," and more particularly the scorbutic diathesis.

From such facts it appears more than probable that ozone is expended in destroying septic impurities.

It is also equally probable that an *excess* of ozone, like undue quantity of very many other substances, is injurious.

The effects of pure oxygen on the respiratory organs is well known ; and, as ozone is oxygen positively electrified, it is not surprising that its issues are still more decided. Thus, when the proportion in the atmosphere was raised from its normal standard of one ten-thousandth part to one two-thousandth, Scoutellen found it sufficiently powerful to kill small animals, and to induce in the human subject powerful respiratory excitation, bronchial spasms, or even inflammation. The *excess* of ozone in the atmosphere has several times been noticed concomitant with epidemics of influenza, complicated with bronchial irritation, and attention has been particularly directed to this subject by Schönbein at Berlin.

[1] Pickford, ' On Hygiene,' p. 70.

Again, the experiments of the authority just named demonstrate that ozone is found during electrical disturbance of the air, and it therefore follows that epidemic disease would diminish after a thunder-storm, which, in popular language, clears the air. By parity of reasoning the same diseases would increase in intensity during hot, close, and heavy weather, and that this is the case we have abundant evidence. For several weeks before the great plague made its appearance in London [1] in 1666, there had been an uninterrupted calm. Diemerbroeck,[2] in giving an account of the plague at Nimeguen, mentions a similar condition of the atmosphere. During the periods of cholera epidemics in Great Britain in 1832, 1849, and 1854, a still and oppressive condition of the atmosphere was observed to prevail. In the last year mentioned, from the 1st of July to the 31st of October, a period of 123 days, a calm was noticed on more than half the number. This was not only the case in England, but was more or less particularly observed throughout Europe, in every locality where cholera appeared. The same have been concomitant on very many occasions in the East, of which the sultry, calm, and clouded atmosphere which prevailed previous to and during the memorable ravages of cholera amongst the troops assembled in Kurrachee, in 1847, is a memorable instance.

The close, still, heavy weather which happened during the cholera epidemic at Paris in 1849; at Berlin in 1855; at Auran, in Switzerland, in 1845; and at Strasburg, was accompanied by the marked absence of ozone.

"Calms," writes Dr. Pickford,[3] "are unquestionably productive of the most baneful and pernicious consequences, by favouring the concentration of miasmata, and of animal and vegetable effluvia, particularly among a crowded and un-

---

[1] Maitland's 'History of London.'
[2] Quoted by Pickford, 'On Hygiene.'
[3] Pickford, 'On Hygiene,' p. 106.

cleanly population, and by delaying those changes in the atmosphere which are necessary to the removal of its purity."

As would naturally follow, periods of epidemic and pestilential disease are associated with a high barometric reading, as was so evident during the cholera epidemics in Great Britain.[1] This atmospheric condition tends to prevent diffusion and aggravate stagnation of miasmatic exhalations. The denser atmosphere sustains more organic impurities than the rarified. The still atmosphere retains miasmata, which air in motion cannot do. Hence the fact of epidemic disease being less frequently prevalent on elevated regions and in the rarified atmosphere of Indian hill ranges than in the dense and more contaminated strata of air resting immediately on the plains below.

The presence of epidemic disease has also been associated with the absence of positive electricity in the air ; and further, the mortality has been found to be in an inverse ratio with the amount of positive electricity in the atmosphere.[2]

It must, however, be recollected that epidemic disease has occurred without *any change having been noticed* in either ozone reactions, in the density or otherwise of the atmosphere, or in the electrical indications which it presents. It would, however, appear certain that deficiency of ozone, a still and dense atmosphere, and a want of positive electricity, are most frequently concomitant, and probably are resulting effects of each other. Further, there is just ground for the belief that one or other of these conditions always occurs during epidemic disease, and that at some period, either previous to or during the ravages of the epidemic, all may be demonstrated.

While, however, these numerous facts lead to the inference that ozone, calms, and electricity are powerful agents in the establishment of health and disease—that their presence or

[1] Glaisher's 'Meteorology.'
[2] Glaisher's 'Meteorology of London,' 1855.

absence favours the diminution or occurrence of zymotic disease;—still, with all such facts before us, we are not justified
in the assertion that any one of the three agents named, or
any complication of the three, is the direct cause of any *particular* class of maladies.

In the same manner, why any peculiar meteorological condition, in conjunction with filth and sanitary defects, shall
cause small-pox in one season or locality, scarlet fever at
another period, or cholera at others, is a complete mystery.
Experience teaches us that with such concomitants some
one or other of these diseases will be originated; but what
subtle combinations are required to induce a particular
malady will probably for ever remain unknown.   It is,
indeed,

> " Causa latet, vis est notissima."

Hence, although we can foretell the origin of erysipelas, surgical fever, irritative fever, hospital gangrene, or puerperal fever
in crowded, ill-ventilated and badly contrived hospitals, still
we are unable to foresee *which* of these particular affections
will become manifest.

Still, however, all these diseases are *ejusdem generis*,
and all exhibit the same class of symptoms in their origin,
progress and termination, whether originating in the temperate or torrid zone.   As Dr. Bennett[1] observes, " This
fever of the country is no other than our own old enemy,
typhus, under a Continental garb.   Its characteristic features
may be modified by some malarious or catarrhal element, but
the type is the same.   The cause, too, is identical in the
Italian marble palace, and in the St. Giles hovel—foul air
inside and outside the house everywhere."

The origin of zymotic disease *de novo* being allowed,
the manner of progress and dissemination becomes the next
question.

[1] Bennet, ' A Second Winter in Italy.'

The simplest kind of contagion is that where a disease is communicated in one way only, namely, by the transmission of a palpable poisonous matter, conveyed from the body of the sick, and applied to the healthy person either immediately or through the medium of anything which has been in contact with the sick, as for example, the bed or clothing: itch, some varieties of porrigo, syphilis, gonorrhœa, hydrophobia, the effects of snake-bite, are in this category.

But there is also every ground for believing that the poisonous matter of certain diseases may be transmitted from the infected to the healthy person, without direct or indirect contact with the palpable virus, and through the agency of the atmosphere. It would appear, also, that the poison passing off, and probably generated in the body of the sick, is identical with that formed externally, and which originated the disease.

In the case of gonorrhœal and other kinds of ophthalmia, it has been frequently remarked, that a visit to a ward where a number of patients are congregated has been followed by an attack of the malady, even when every care has been taken to avoid coming in contact with the sick, or with any articles near them.

Thus, latterly in the Orphan Asylum, near Prague, an epidemic of ophthalmia broke out, and ninety-two children were attacked. M. Eiselt examined the air with Pouchet's acroscope, and, in the atmosphere of a ward where lay a great number of children, a quantity of large pus-cells were found. The cells were noticed on the instrument immediately the air was made to pass through the apparatus.

In such cases, it may be probable, that the pus-globules floating in the air attach themselves to the eyes of healthy individuals, and so excite the disease; but there are other affections which arise from the atmospherical dispersion of the *materies morbi*, and its consequent absorption by the skin, lungs, or alimentary passages. Cholera, yellow fever,

malarious fever, influenza, small-pox, plague, and the whole
of the exanthemata, obviously belong to this class.

Some of the diseases mentioned, however, are communicable in both ways, viz., by direct contact or application,
and also by atmospheric dispersion of *fomites* or *materies
morbi*. Of these, small-pox—(according to Mr. Youatt—
glanders) and puerperal fever are examples.

It is frequently a matter of surprise that individuals are
stricken with disease of the zymotic variety who have apparently never been exposed to contagious or infectious influence, and who live without any great defect of sanitation
being discernible in the locality of their residence. As was
demonstrated, however, during the Windsor epidemic in
1858, disease will originate if its causes accumulate, even in
the proximity of the most stately residence. At that
period, as has been remembered with suspicion[1] since the
melancholy death of the Prince Consort, Dr. Murchison
showed, that only one half of the Royal Mews participated
in the efficient system of drainage applied to the castle.
That part which was well drained, remained free from fever ;
in the other portion, as in the insufficiently drained town of
Windsor itself, disease was rife, thirty cases of fever and
three deaths occurring among the royal servants.

The question naturally arises, at what distance from the
bodies of the sick, or from the localities of formation, is the
miasm capable of inducing disease? This can scarcely be
reduced to demonstration. Both classes of emanations become fatal, or prove innocuous, according to the extent to
which they are diluted by pure air. Miasm may be, roughly
speaking, said to diminish as the square of the distance ;
and, in fresh air, according to Dr. Alison, it does not extend
beyond three feet from the patient. When, therefore, fever,
erysipelas, or gangrene, spreads in an hospital, or originates
in its wards, such an event proves that the laws of nature are

---

[1] 'Lancet,' Nov. 21st, 1861.

not being carried out—that the miasms from the sick have so injured and dirtied the air, that it is unfit for the respiration of human beings. If, however, the same air is but partially soiled, it can only operate with a less degree of force. Therefore, when we take into consideration the prevalence of draughts of air in various directions, particularly among buildings, and the probability of the presence in the atmosphere of various gaseous substances, which, like ozone, may have the power of destroying *materies morbi*, it ceases to be surprising that we cannot define the exact limits at which miasm loses the power of exciting disease. Indeed, as previously stated, a small dose of the poison, that is to say, breathing less dirty air, will, perhaps, not induce the typhus or hospital fever from which it arises, but will cause headache, lassitude, or *malaise* instead.

Moreover, much must depend on the state and condition of individuals. A weakened or fatigued person will become affected more easily than a robust and vigorous man ; the drunkard sooner than the temperate ; and the hungry and poverty-stricken before the well-fed and opulent. Also season exerts an influence as Dr. Smith[1] shows, depressing the vital powers to a minimum during the autumnal period, or just at the time when the cyclical change of the year induces the formation of the largest amount of *materies morbi*.

Hence it cannot be stated with any exactness how near an individual may approach the poisonous locality (whether such locality be the bodies of the sick or other generators of miasmata), without running danger of becoming affected ; but we can in this manner account for the fact of individuals contracting disease who unknowingly inspire the poisoned atmosphere from an open door, an untrapped drain, or other locality, which, perhaps, one moment afterwards, in consequence of a current of air sweeping by, may be healthy for the next and more robust passenger.

[1] Smith, 'Cyclical Changes,' p. 161.

It is on this principle that a windward position is always more healthy than the reverse. And it is this knowledge which enables the medical man so frequently to escape disease when visiting and examining the infected sick. If there be any draught, the physician almost instinctively passes to windward of his patient, and before examination of the back, for instance, would allow time for a circulation of air between his patient's person and the bedding.

It is also a rule worth remembering and acting upon, to clear the mouth and expectorate, after exposure to any probable source of infection, by which means there can be little doubt that miasm may oftentimes be rejected, before it can enter the system.

The researches of Schröder establish beyond doubt that the formation of the hyphaceous fungi previously referred to will not take place in a bottle containing organic matter previously heated to boiling, provided the neck of the flask be loosely plugged with *cotton*. The substance remains for years unaltered, fermentation and putrefaction being arrested by the cotton acting as an air-filter, and retaining the solid particles suspended in the atmosphere. The later experiments of Pasteur "on spontaneous generation," also led to the same results. Hence the value of a respirator to those exposed to soiled air.

To conclude, experience has incontestibly shown, that just as disease never springs up primarily in a place where the atmosphere is fresh and pure, so it cannot continue to exist, far less to thrive, where such an atmosphere prevails. "The same aerial condition is necessary for the reproduction as for the original production of the morbific germ."

CHAPTER III.

ON MALARIA.

SEVERAL authors have lately attempted to prove that the agent we call malaria has no existence, and that the paroxysmal diseases generally associated with such miasmata are due to other causes. It is not difficult, in the absence of visible evidence, to give the apparent colouring of truth to any stated proposition, especially when the demonstration of the subject depends on analogous reasoning, instead of direct testimony ; but, on the other hand, it is comparatively easy to prove to all "who see but what they list" that certain events invariably following stated antecedents must have their origin in the latter.

The theory of the author of the Brownonian[1] system of medicine, which, about the close of the last century, gained so great an ascendancy in both professional and public estimation, "that debility is the cause of intermittent, as well as of every other kind of fever," appears to be again obtaining supporters.

[2] Brown, ' Elementa Medicinæ,' prop. xxii.

Without referring to the opinions of the late Dr. Todd, whose heroic stimulating method of treatment bears on this subject, or to the ideas of those who support that lamented physician's views, it will be sufficient to allude to the statements of authors who have *limited* their remarks to malarious disease.

Thus, Dr. Smart[1] "is disposed to consider that the fevers of Hong Kong must not be brought forward to swell the charge of insalubrity of climate," and remarks "that, in the majority of fever attacks at that place, it will be ascertained that simple exposure to the sun, disregard of sanitary precautions, either voluntary or enforced, indulgence in habits that derange the digestive functions or depress the nervous powers, have preceded the invasion of fever, standing in the relation of exciting causes."

Again, Dr. Burdel[2] regards marsh poison as a "myth," and looks upon malarious fever as a "perturbation of the cerebro-spinal centres and of the sympathetic system"— adopting nearly the same nomenclature as the one by which Bernard defines "glycosuria" (with which also Dr. Burdel connects paroxysmal disease), but at the same time forgetting to explain the reason of this "perturbation" of the nervous system.

Another gentleman, Dr. Knapp,[3] President of the University of Iowa, brings forward a remarkably easy doctrine— allied to the homœopathic absurdity—viz., that the "scorbutic diathesis is primary pathology," and that all known causes of pestilence produce their results by impairing the nutritive functions, and thus does not admit any hypothetical causes, such as "supposed malaria, epidemic influences, or occult qualities," which terms the doctor states are only "cloaks for ignorance, and hinder the progress of science."

[1] Smart, "On the Diseases of Hong Kong," 'Lancet,' Aug., 1861.

[2] Burdel, 'L'Union Medicale,' No. 139, 1859.

[3] Knapp, 'Researches on Primary Pathology and Laws of Epidemics.'

That the scorbutic diathesis is a powerful predisposing cause of disease, will be sufficiently demonstrated in Chapter XX, "On Scurvy;" but that intermittent or paroxysmal affections will arise from this alone cannot be correct, or we should find malarious disease prevalent amongst Arctic voyagers, which, however, is not the case.

Linnæus,[1] in his thesis entitled 'Hypothesis Nova Febrium Intermittentium causa,' announced the true cause of ague to be 'aqua scilicet argillacea,' or the use of water impregnated with inorganic matter, while percolating through an argillaceous soil. That impure water induces malarious disease will be shown in the chapter devoted to a consideration of that fluid; but the cause of this would appear to be, saturation of such water by malaria, and not the admixture of any other agents, which, as demonstrated in the chapter just referred to, excite diseases of a widely different nature.

It will now be shown how far facts agree with the assertions or opinions of the authors quoted.

It has already been advanced (Chapter II), that the most skilful analytical chemists have been unable to detect any alteration in the composition of the atmosphere, whether the air submitted to analysis was procured from mountains or plains, cities or deserts. It has also been shown that the peculiar decomposition originating any particular disease has not yet been ascertained. In like manner, the determination of the precise nature of malaria has hitherto eluded discovery, although both chemists and microscopists have devoted much time and attention to the investigation.

Dr. Pickford[2] states :—" Notwithstanding our ignorance of its chemical and physical properties, there is every reason to believe that malaria is an organic compound, chiefly composed of carbon and hydrogen."

According to Schönbein[3] marsh miasm consists of gaseous

[1] 'Amœnitates Academicæ,' vol. x.
[2] Pickford, 'On Hygiene,' p. 166.
[3] Schönbein, ' Med.-Chirur. Trans.' vol. xxiv.

matters, the products of purely chemical or physical or
physiologically-chemical action, taking place within the earth
or upon its surface, in stagnant water, or in the atmosphere
itself.

Galen, Arctæus, Celsus, and all the older authors, associated malaria with swamps and marshes; but it is now thoroughly proved that, as Dr. Maculloch stated, "before
malaria can be produced, it is necessary that the land should
be visible and drying," or otherwise subject to peculiar alternations of dryness and moisture. When a swamp is covered
with water, no malaria is extricated. It appears probable,
from instances similar to those quoted in the chapter on
water, that this fluid *absorbs* the malaria, and thus prevents
its dissemination through the atmosphere. We cannot otherwise account for the fact of malarious fever following the use
of stagnant water as a drink. The absorption of malaria by
water also accounts for the idea that malaria cannot pass over
the latter fluid; although, indeed, the greater amount of
ozone in the atmosphere above lakes and rivers must also
have its share in destroying or neutralising the poison.

Hence malaria chiefly abounds on the margins of swamps,
or where the soil is boggy and drying; on plains which have
been flooded; on alluvial shores; on the deltas and in the
course of tidal rivers which are twice in twenty-four hours
influenced by the ocean; in the dry beds of tropical rivers;
on plains and level countries presenting physical obstacles to
drainage; in the rocky hollows and alluvial soils of mountain
valleys; and in all soils generally which afford capabilities
for the retention of moisture.

A malarious soil is indicated by marshy vegetation, and the
presence of amphibious animals of the Batrachian kind, such
as frogs and tortoises. The presence of innumerable flies, of
all shades and colour, locusts, insect-eating birds, birds of
the genus Scolopax—as was the case in the "valley of death"
where our troops suffered so severely—point out an unhealthy
malarious climate.

But not only is malaria extricated from marshy ground, but it is also produced on sandy alluvial plains, where little or no vegetation exists; as at Walcheren, where 18,000 men died or were invalided in three months; at Baia, where a French army of 28,000 men were reduced in a few days to 4000; in Spain, on the plains of Estremadura; at Ciudad Rodrigo; and in very many tropical districts.

Again, Sir Ranald Martin[1] has shown that many localities in India, where ferruginous soil or hornblende persist, are essentially productive of malarious disease. Sir Ranald has also demonstrated, by the aid of Sir Charles Trevelyan, that at Hong Kong, Sierra Leone, and other places, the same affinity exists between ferruginous soils and malarious disease. To this list, in my ' Report on the Medical Topography of Bhooj,'[2] I added the province of Kertch, in Western India. Dr. Heyne,[3] in his paper on the ' Hill Fevers of the South of India,' also dwells on the connection which appears to exist between ferruginous earth and malaria.

It may, however, be asserted that all soils exhale malaria, some, probably, to a larger extent than others. The reason why miasm is more powerful in particular localities has now to be explained.

It appears extremely likely that, as Dr. Parker[4] holds, malaria does not arise from any cause existing *on* the surface of the ground, but that it depends on chemical or volcanic action occurring *beneath* the surface. Otherwise, how is the presence of malaria to be explained in those districts which are arid, and devoid of vegetation? It would seem, however, that the formation of malaria does not take place at any great depth below the ground, *otherwise colliers and miners would be subject to intermittent disease, which certainly is not*

---

[1] Martin, ' On Tropical Climates.'
[2] ' Bombay Med. and Phy. Soc. Transactions,' 1861.
[3] ' Indian An. Med. Science,' vol. i.
[4] Parker, ' Causation and Prevention of Disease.'

3

the case. Malaria, however, being extricated, becomes entangled in the damp vapours, air, and rotting vegetations of swamps, which will *account for their pestiferous character.*

Dr. Hirsch[1] has latterly brought forward several arguments in opposition to the view previously entertained, that malaria is the product of decomposing vegetable matter, and instances several localities where there are moist swamps, apparently presenting all the climatic and terrestrial characters of other regions where malarious fevers abound, but which are free from such diseases. These localities are several places in Peru, the Pampas of Rio de la Plata, some positions on the delta of the Mississippi, and Negapatam in the Madras Presidency. Regarding the latter place, however, I have reason to believe that malarious disease exists there, as in most places in India. Yet it is an acknowledged fact, that peat bogs do not emit malaria, and also that the Dismal Swamp, a moist tract of 150,000 acres, on the frontiers of Virginia and Carolina, does not induce aguish fever.

Hence it would appear that in such places, and also elsewhere, there must be some counteracting influence at work, preventing either the generation or extrication of miasmata, or destroying it immediately on formation. Analogous reasoning will not allow us to suppose that localities having identical similar climatic and other characteristics do not obey the same laws, and therefore it cannot be imagined that such surfaces as the Dismal Swamp soil do not *generate* malaria. It is, however, consonant with our knowledge of the mercy and bounty of Providence to suppose that, as He has appointed organic life to consume the expired carbon of animals, and the reverse, so in like manner vegetation may be made in other ways the cleanser and purifier of the atmosphere, and this, indeed, science gives us cogent reasons for asserting is the case.

[1] Hirsch's 'Handbook,' p. 67.

Pliny long since maintained, that groves and trees absorb and destroy mephitic vapours.

Dr. Kinloch Forbes holds that trees live on miasm; but even if they simply decompose it and absorb the carbon, which appears to be one of its constituents, in the same manner as they decompose and absorb carbonic acid, the beneficial effect—the destruction of malaria—is realised. It has also been supposed that trees prove antagonistic to malaria in another manner, viz., that the oxygen they exhale into the atmosphere becomes positively electrified or, in other words, converted into ozone, the disinfectant powers of which have been already considered.

That trees act in one, probably in both, of these methods on malaria, that they respire or absorb mephitic carbonaceous emanations from the earth's surface, and that the oxygen they so plentifully form becomes converted into ozone, for the more perfect destruction of marsh miasm, there are many convincing arguments.

Malaria is most pestiferous and active from sunset to sunrise, when plants respire oxygen, and give off carbonic acid; when, in fact, they cease to absorb the latter, and, probably, the carbon of malaria.

Malaria has most effect on the system during the hours of sleep; this obviously depending partly on its not being destroyed by night, and its consequent greater concentration and power, and partly on the known fact of the human system being less able to withstand subtle causes of disease when in the weary state, demanding sound repose. As noticed in chapter XII, "On Clearing," natives of India dread sleeping under trees having dense foliage, as the nimb and the tamarind.

It is stated that several localities in America, formerly malarious, have been made the reverse by planting sunflowers, which there attain the greatest bloom and vigour during the autumnal season, or at that period when most vegetation decays, and the largest quantity of malaria is extracted. Lieut. Maury, of the American navy, appears to have been the first to direct attention to the anti-malarious properties of the sunflower.[1]

The effects of malaria are, for the most part, produced in autumn, when vegetation is decaying and drying up, and therefore ceases to absorb carbon in the same quantities as before.

Swampy localities not productive of malaria, as the Dismal Swamp, the delta of the Mississippi, &c., are covered with evergreen trees, which retain their foliage throughout the entire year. Peat bogs are constantly green.

In hot climates, where malaria abounds, most of the trees are evergreens.

A screen of trees between a swamp and town has rendered the latter healthy; removal of groves the reverse.

Lastly, malarious fevers have been observed especially virulent where no vegetation existed.

Moisture and a somewhat elevated temperature are necessary to the development of malaria. This process, like putrefaction, is suspended by perfect dryness, and a temperature below freezing point; also, to some extent, by the scorching heat of a summer tropical sun. Hence paroxysmal disease does not exist in the Artic regions, and not nearly to the same extent in the dry seasons of tropical climates.

Malaria, in its diluted forms, gives rise to various anomalous symptoms, which have been latterly referred to by Dr. Handfield Jones, in his brochure on 'Neurolytic or Aguish Disorders prevalent in London.' In tropical climates the saturation of the system by malaria is much more decidedly marked, and is the chief cause of those conditions described in my 'Manual of the Diseases of India,' under the heads of "Cachexia Loci," "Splenic Leucocythæmia," and "Masked Malarious Fever."

[1] 'Western Lancet,' 1860.

In fact, malarious influences are the chief causes of that degeneration of blood and *vis vitæ* which commences from the day the European sets his foot on a tropical or semitropical shore; and every disease in such countries is impressed either in its origin, progress, or termination, with the paroxysmal phenomena of the malarious taint.

The malarious type of dysentery, is as Deputy-Inspector-General Cameron[1] truly observes, one of its most fatal forms, and " must be regarded, in a great measure, as essentially fever."

Some also have supposed that ardent continued fever, plague, pestilence, cholera, and epidemics of heat, asphyxia, are the results of malaria.

Although there are arguments in favour of malaria being the exciting cause of the terrible diseases just named, still it appears most probable that malaria, as is the case with dysentery, merely aggravates their symptoms. As Mr. Murray[2] states of cholera, " Many of these cases towards the end of the epidemic assumed an intermittent form, with profuse perspirations, which modification was evidently caused by the miasmatous atmosphere ;" so it is found in all malarious countries that epidemics of any disease are influenced by the atmosphere acting on a weakened constitution, and exciting those manifestations which it could not do in a state of health.

In addition, however, to the malarious influence excited over epidemic disease, the same taint is recognised in other and widely different ailments. Thus, Dr. Francis[3] records several cases of gout having a malarious origin ; Dr. Geddes[4] mentions convulsions as arising from miasm ; and Dr. Payne[5] has recently published an ' Essay on Epileptiform Seizures'

[1] Cameron, " On the Employment of Quinine in Fevers." ' Lancet,' 1861.
[2] Murray, " Report on Cholera in India." ' Ind. An. Med. Sci.,' vol. xiv.
[3] Francis, ' Ind. An. Med. Sci.,' vol. xiv.
[4] Geddes, ' Clinical Illust. of Disease in India.'
[5] Payne, ' Ind. An. Med. Sci.,' vol. xiv.

dependent on malaria. Several authors have also described convulsions occurring to children in India as due to the same causes. In a more intense form, or in weakened or debilitated habits, or in the autumn season, when, in addition to the greater intensity of malaria from causes already explained, the human system is at the minimum as regards muscular and vital power, malaria gives rise to ague or intermittent fever. In cold countries, where the malaria is not so abundant, the tertian type is generally observed; in the tropics, from the causes previously referred to, quotidian fever generally occurs. In still greater intensity malaria excites remittent fever, with all its complications and tendencies to become continued and typhoid.

Malaria, however, may be so intense as to kill at once, as has occurred on many occasions both in semi-tropical and tropical climates, as in Italy,[1] India, and in the West Indies.[2]

In other instances bowel complaints, as the "dry colic" of the West Indies, the dhobee's colic of the Indian hill stations, and common diarrhœa, are induced by exposure to malaria, *the poison appearing to expend itself in the flux induced.*" "Dhobees," or washermen, frequently suffer from colic and diarrhœa, after remaining long in the narrow ravines, where they proceed for the sake of water, and sportsmen are often attacked by diarrhœa after wandering through malarious localities.

" Fever," says Dr. Maculloch, "may be induced within half an hour after exposure to malaria;" the more immediate effects, as above demonstrated, in even less time. Again, malarious fever may occur at sea, as happened in my own charge to individuals who had lost sight of land for ten or twelve days, *and who had never before suffered from intermittent.*

"Malaria loves the ground," and therefore is found in greater intensity in valleys, which accounts for some of the

---

[1] Pickford, 'On Hygiene.'

[2] Fergusson, "On Marsh Poison;" 'Trans. Ed. Roy. Soc.,' vol. ix.

facts previously stated, and also is the reason why, in malarious countries, the inhabitants of ground floors are affected
by disease in a greater proportion than those sleeping in
upper apartments, as was painfully demonstrated in the
hospital at Padua,[1] at Guadaloupe,[2] at Seringapatam,[3] at
Calcutta,[4] in 1833, when the fort ditch was allowed to empty
itself, and the inhabitants of lower storeys suffered in consequence, and at other places.

It has already been stated that malaria may be entangled
in fogs and mists, and hence it is the more readily movable
by the winds, although indeed it may be so extended without
the aid of damps as a vehicle. Thus Lancisi[5] mentionsthat
out of a party of thirty, who had sailed to the mouth of the
Tiber, twenty-nine were attacked with ague on the wind
shifting and blowing over a marsh to windward.

In America, Holland, India, and other countries, much
the same results have been recorded.

In consequence of the preceding facts, all writers on malaria, especially Drs. Fergusson, Corbyn, and Mackinnon,
concur in insisting upon the evil effects of choosing a position
to leeward, or within or at the mouth of a ravine, in the
lowest depths of which, rarely thoroughly perflated in still
weather, or visited by the direct rays of the sun, miasms
stagnate, and from which they roll in a most concentrated
form. Thus it is concluded that the specific gravity of malaria is greater than that of air.

It has already been demonstrated that trees decompose or
absorb malaria, and they must, therefore, in common with
any other obstacles, as a high building, arrest the poison in
its course. Unlike masonry, however, they *retain* it, and do
not admit it to blow through or over their foliage—at all

---

[1] Clarke, 'On Climate.'
[2] Fergusson, "On Yellow Fever;" 'Med.-Chir. Rev.,' vol. xiii.
[3] Ballingall.
[4] 'Ind. An. Med. Sci.,' vol. i.
[5] Lancisi, ' De Noxiis Paludum Effluviis.'

events to any considerable extent. This fact has been frequently turned to practical account in India and elsewhere. Many places have been made comparatively healthy by planting a screen of trees between them and a malarious locality; while, on the other hand, many salubrious places have become the reverse, by destruction of their vegetable safeguards.

"A forest," says Mr. Haviland,[1] "intervening between a pestilential marsh and a city, often affords protection to the inhabitants;" and Fergusson[2] wrote that, in the territory of Guiana, "It is wonderful to see how near to leeward of the most pestiferous marshes the settlers, provided they have the security of trees, may venture with safety."

Cutting down screens of trees has ever been a fatal operation, as was demonstrated at Paramariboo, the capital of Surinam; in the Campagna of Rome; at the Convent of St. Stephano, and elsewhere.[3]

But although vegetation and trees are beneficial in the most extended sense, they may be made just the reverse, if any one perversely camps or resides under their shade when, from the character of the adjacent surface, there is reason to believe the accumulation of miasm may take place in their foliage.

Nothing checks the generation and propagation of malaria so much as dense population and cultivation. Even, however, in the latter salutary process there is a certain amount of danger. Inefficient drainage may just bring a swamp into that drying and boggy condition which is so peculiarly favorable to the extrication of malaria. This was the case after the partial draining of the marsh of Chartreuse.[4] Again, too profuse irrigation may induce disease; and hence the ob-

---

[1] Haviland, 'Climate, Weather, and Disease,' p. 58.
[2] Fergusson, "On the Nature and History of Marsh Poison;" 'Trans. Roy. Soc. Edinb.,' vol ix.
[3] Pickford, 'Hygiene;' Art. "Malaria."
[4] Martin, 'On Tropical Climates.'

jection which Col. Baird Smith[1] states has been taken in Italy to rice cultivation. Indeed, it is forbidden in that country to establish these "marcite" meadows within 1000 English yards of forts or fortified places, within three miles of the communes, and within five miles of the capital. It would be well if these regulations were made applicable to the irrigated land so frequently too near our Indian cantonments.

Again, the first cultivation of a district, the first ploughing up of the land, is said to be always productive of malaria.[2]

It has been stated that population tends to check the formation of malaria; and this arises in two ways—first, from the cultivation which a populated country undergoes, and, secondly, from the number of fires which population and civilisation demand. It is stated that the postmaster of a station in the Pontine Marshes preserved his health by keeping a large fire constantly burning day and night throughout the year. That fires, or their products, disseminated in the atmosphere, will destroy the epidemic constitutions of the air during times of plague, we have abundant proof, both from ancient and modern history. Thus Hippocrates and Pliny state that the Athenians were cured of the plague by lighting fires near the houses of the sick. The same disease disappeared after the great fire of London, in 1666. The large manufacturing cities of Manchester and Birmingham passed through the cholera period almost untouched; the affection was arrested at Varna after the conflagration of the town. There can, therefore, be no doubt *that fire exerts a destructive influence on malaria,* as over the subtle atmospheric causes of disease. Hence the desirability of lighting fires round encampments placed in a malarious locality.

In consequence of malaria being so much more powerful in valleys and hollows than on elevated sites (a fact which has been instinctively taken advantage of by the inhabitants of

---

[1] Smith, 'Italian Irrigation.'
[2] Pickford, 'Hygiene;' Art. "Malaria."

malarious countries, as in Sicily, and many parts of India, where the villages are frequently placed on hills), it has been thought that at a still greater elevation it could have no existence. Hence it was fondly expected that our Indian hill stations would remove the dwellers thereon above the range of malarious disease. Unless, however, the elevation is sufficient to bring the inhabitants of such localities near the line of eternal snow, *the concomitants of alluvial soil in ravines, periodical rains, and a vivid sun will not fail in inducing the extrication of malaria*, which, although *less powerful* than that generated on the plains below, will still induce its characteristic manifestations. (See chapter on Hill Sanitaria).

Lastly, people born and bred in a malarial country appear to become inured to it, and enjoy comparatively good health, where new-comers would suffer. The Garrows, on the frontiers of Assam, are a powerful race, but inhabit a country into which no European could penetrate without the almost certainty of contracting a dangerous fever[1], and the slaves of the Confederal states of America find safe asylums in deadly jungles, knowing that their masters cannot follow them there without grave risk of life.

Practically, our knowledge of malaria leads us to avoid marshes, fens, moors, valleys, low-lying ground, ferruginous soils, and sandy wastes, the banks of tidal rivers, the course of dried-up streams, and localities destitute of vegetation, as encamping ground for troops, or as sites for cantonments, &c.; also, when *obliged* to occupy such localities, to pass to windward, to obtain a screen of trees, not to reside beneath trees, to live in upper buildings, to keep doors and windows, and tent "purdahs" closed, particularly those facing malarious localities; to avoid the latter at night; if exposure be obligatory, to fortify the system by a previous meal, with a small supply of fermented liquor, to use a respirator, or, perhaps, smoke tobacco.

---

[1] McAsh, 'Advice to Officers in India.'

CHAPTER IV.

SHOWING THAT IN ALL AGES SANITATION HAS BEEN REQUIRED, AND ITS NEGLECT FOLLOWED BY DISEASE.

Epidemics of the Middle Ages—Leviticus—Hindoos and Mahometans—The Gaol Fever—Mahammurree—Dysentery—Zymotic Disease: causes not understood—Conveyance of Disease to Army and Navy—Condition of the Soldier some years since: his Barracks; his Diseases; his Hospitals—Deductions from the foregoing.

To prove this, it will not be necessary to detail the various accounts of ancient epidemics which are extant. It must be admitted that defective sanitation thousands of years since will have given rise to much the same diseases which it allows to occur at the present period; otherwise, probably, Hecker's 'Epidemics of the Middle Ages,' and the account of the plagues of the sixth century which Gregory of Tours has recorded, will be satisfactory proofs of the foregoing remark.

It may be supposed that even the most sceptical will not deny that many of the laws promulgated by the inspired Moses were sanitary regulations in their fullest acceptance, intended for the enforcement of cleanliness and, thereby, the removal of the causes of zymotic disease from the Israelitish camps.

Among these may be mentioned, the directions what to eat and avoid (Leviticus, chap. vii, chap. xi); the purification of women (chap. xii); the laws concerning leprosy (chap. xiii); the disposal of excrete (chap. xv), &c.

Had it not been for such wise regulations, we of the present

day, with our knowledge of the ravages of epidemics in later
times, have no difficulty in surmising the diseases which
would, in a semi-tropical climate, have decimated the camps
of the children of Israel.

Again, we may trace sanitary regulations in very many of
the religious ceremonies of the people among whom we live—
the Mahometans and Hindoos of the East. Were certain
regulations of the above religions nullified, and the *general
meagre* sanitation of European civilisation *not* substituted,
the cities of India would become infinitely more prolific in
the production of zymotic disease than is at present the case.

From among the numerous examples which sanitary
history presents, both in ancient and modern times, of the
danger which results from neglecting the principles of health,
it is almost difficult to choose. Perhaps the history of the
gaol fever in England is as "apropos" as any other occurrence of the kind.

When we recollect the conditions of prisons and gaols, as
described by Pringle[1], by Mead[2], by Howard, in his great
work, and by Lord Bacon; when we recollect that such
buildings were ill-contrived, frightfully crowded, that conservancy was not practised in the slightest degree, that the diet
was of the most meagre description, and that the operation
of a sanguinary and ill-regulated penal code produced a corresponding mental depression; when we recollect all these
things, it ceases to be a matter of surprise that, as Mead
wrote, "Very few escape the gaol fever, which is always
attended with a degree of malignancy in proportion to the
closeness and stench of the place."

The first outbreak of gaol fever of which we have an
authentic account, took place in 1414, when the prisons of
Newgate and Highgate were severely visited. In 1522, an
outbreak took place at the Session Hall, in the Castle of

[1] Pringle, ' Observations on the Gaol or Hospital Fever.'
[2] Mead, ' Discourse concerning Pestilential Contagion.'

Cambridge. This was followed, in 1577, by the celebrated black assize at Oxford, when 510 persons were infected and died. A similar outbreak to that at Oxford, took place at Exeter, in 1585. Gaol fever was found to prevail in the 18th century, whenever from any cause gaols become crowded; as in 1730, from the large amount of crime; in 1749, on the conclusion of the peace of Aix la Chapelle; and under similar circumstances in 1783.

The sickness which originated in these over-crowded gaols was not, however, by any means confined to the inmates of such buildings. The disease spread not only into the court-houses in which the prisoners were tried, but also even beyond those localities.

Knowing what we now do of the cause and dissemination of disease, it ceases to be a matter of surprise that at the Old Bailey, in 1750, an outbreak occurred, when, curiously enough, as then thought, those to the right of the Lord Mayor generally escaped; the reason being, that a stream of air from the window directed the *materies morbi* to the other side of the court.

Neither can we wonder that as Baker relates, at the black assize at Oxford, July, 1577, " Suddenly they were surprised with a pestilent savour, whether rising from the noysome smell of the prisoners or from the damp of the ground is uncertain, but all that were present, almost every one, within forty hours died. There died Robert Rich, Lord Chief Baron, Robert D'Oylie, Sir William Babington, &c., &c., almost all the jurors, and 300 others, more or less "! Or that, in 1588, an infectious distemper destroyed the judge and most of the persons summoned to the Lent assizes at Exeter. That in 1730, at Taunton, Chief Baron Pengally, with his officers and servants, died, having taken an infectious disease from the prisons. That in 1742 a putrid fever appeared at Launceston, generated by the prisoners. That on May 11th, 1750, at the sessions at the

Old Bailey, four on the bench, two or three of the counsel,
an under-sheriff, several of the Middlesex jury, and others
to the number of forty, were attacked with putrid fever, and
perished, with many other instances of similar nature, which
might be quoted.

A parallel to the gaol fever may be found in the prevalence
of "mahammurree" in the Himmalayahs, the history of which
will be presently referred to.   There the inhabitants, not
attending to personal cleanliness, and in houses where the
pure cold air is shut out as an enemy, and every crevice made
as impervious as possible, have been wont to sacrifice health
for warmth, endeavouring spontaneously, as the prisoners in
English jails were formerly obliged to do, to live in defiance
of every sanitary law, and paying the certain penalty, in the
ravages of a disease of similar nature.

It is well known that dysentery is a well-marked instance
of a disease which, not primarily communicable, is apt to
become apparently contagious under certain circumstances
favorable to its existence.   Thus we find that whereas the
mortality from this disease does not now in England and
Wales exceed from 20 to 100, in the eighteenth century it
averaged from 1000 to 2000 annually.

We seldom now hear of persons in comfortable circumstances dying of dysentery in the United Kingdom ; but Dr.
Chevas[1] enumerates a number of notable persons who died
of this disease.   Among them are the names of Prince Henry,
son of Henry II ; Edward I ; Cardinal Wolsey ; Robert
Dudley, Earl of Leicester ; Walter, Earl of Essex ; Henry
VIII ; James I ; Oliver Cromwell ; Charles II, &c.   The
dysentery in some instances was accompanied by malarious
fever.

The Report of the Registrar-General for 1859, contrasts the
number of deaths from zymotic disease, occurring during the
period from 1810 to 1859, with the mortality from 1660 to

[1] 'Ind. An. Med. Sci.,' vol. xiii.

1679; by which it appears that an individual was then in four times more danger of dying than he is now from such affections.

Before the publication of the works of Pringle, Howard, Lind, and others in the last century, the true causes of such epidemics and their removal were but imperfectly understood.

Hence, as the army and navy were in those days generally recruited from the gaols, it not unfrequently happened that the "sickness of the house" was conveyed to regiments and ships' companies. Indeed, the writings of Pringle, Bancroft, and Lind, demonstrate that hospital, barrack, and ship fevers, were precisely of the same type as that so prevalent in gaols.

Dr. Lind, in his 'Essays on the Health of Seamen,' wrote:— "The source of infection to our armies and fleets is undoubtedly the jails ; we can often trace the importers of it directly from them. It often proves fatal on impressing men, on the hasty equipment of a fleet. The first English fleet sent last war to America lost by it above two thousand men." The seeds of infection were carried from the guard-ships into our squadrons, and the mortality thence occasioned *was greater than by all other diseases or means of death put together.*

It was not, however, conveyance of disease from gaols, which *alone* caused the mortality amongst the army and navy in those days.

Dr. Chevas[1] states :—"It was not until the first half of the eighteenth century was past that England could reconcile herself to a standing army. About 1739, several barracks were built in the neighbourhood of London, an innovation which excited a great deal of angry discussion in the public mind. These barracks were a sight to which this nation hath not been accustomed. They had very much the air of

---

[1] 'Ind. An. Med. Sci.,' vol. xiii.

those citadels which in arbitrary countries are erected to keep the people in awe. The citizens of London took the alarm, and cried out with one voice, they would have no red-coat nurses. Indeed, one popular writer of the day was of opinion that if it had happened in the days of our ancestors, the director would have had very good luck if he had died in his bed."

We may, therefore, readily expect to find that the comfort or health of the troops was not considered.

In many of the barrack-rooms the men had to sleep in bunks, which were places arranged like the berths of a ship, one tier over the other. To each tier were told off four men (and they were hardly large enough to contain three), as it was supposed one man would be on guard or duty, so that three men only would have to sleep in it. These rooms were frequently over stables. Only imagine three or four men sleeping in the same bed, and these beds one above the other, not more than two feet between each, always two, sometimes more than three tiers high. Though this is hardly to be credited now, yet it is a fact; and a room formerly used for this purpose may be seen at Woolwich, and is in the east square, on the left-hand side of the archway going out.[1]

What must this have been, if one or two drunken men got into the upper bunk—what a treat for a steady sober man, to have to lie between two drunkards, men or women, the latter of whom were also frequently brought into barracks.

We cannot, therefore, wonder at the statement made by Dr. Brocklesbury, physician to the army, who, in 1763, declared "that the register kept of the military who died of fever of various kinds proves that more than eight times as many fall by the petechial or jail fever, as by battle."

If such diseases originated in barracks and gaols, how much

[1] 'United Service Magazine;' "The Soldier of Fifty Years ago," by Major Shaw.

more rife must they have been in the hospitals of the period, where wounds and putrefying sores abounded, and where the disgusting custom of placing more than one patient in each hospital bed was usual.

Dr. Trotter[1] states that, at the sick quarters at Dartmouth, men with fevers were found two and three in a bed; and not only was this the case in military hospitals, but also a common practice in civil institutions of the period, both in England and on the Continent.

In short, those who are aware of the former state of our barracks, hospitals, ships, gaols, and all public institutions during the earlier part of the last century, and previous to that period, and who are able to compare such condition with the existing state of things, must allow the truth of the heading of this chapter, *in its most extended signification;* and more than this, they must also feel certain that extension of the present sanitary system will eventually, as stated in Chapter I, exterminate the whole class of zymotic disease.

---

[1] 'Medicina Nautica,' vol. iii.

CHAPTER V.

Allusions to former Mortality of Europeans in India—Recent Mortality—A
Contrast—Invaliding—Expense of Disease in India—Mortality of Women;
of Infants and Children.

Sir RANALD MARTIN[1] quotes numerous instances from the writings of older authors, showing the general unhealthiness of Calcutta. In addition to this, it is elsewhere stated[2] that, of the troops which Sir Abraham Shipman brought with him to Bombay, in 1662, there remained in 1664 only 93 out of 500. Dr. Fryer, who visited Bombay in 1672, said of the Europeans there, "I reckon they walk in charnel-houses. In five hundred 100 survive not." Mr. Ives tells us that, at the time of Major Kilpatrick's death, in 1757, only five of the 250 soldiers who accompanied him from Madras in August of the previous year survived him. Captain Hamilton, who traded in India between 1688 and 1723, makes frequent allusions to the unhealthiness of Europeans, and Dr. John Clarke, who was in Calcutta in 1768, records that, out of 189 cases of fever treated in ships, only 84 recovered!

That the mortality *during the last few years* has been enormous in tropical climates is sufficiently evidenced by the fact that the death-rate of European soldiers in India, as nearly as it can be ascertained from the most available data,

[1] Martin, 'On Tropical Climates,' p. 141.
[2] Chevas, 'Ind. An. Med. Science,' vol. xiii.

has been, *excluding all sudden casualties, out of hospital,* as
suicide, accident, and killed in action—

62·45 annually in the thousand.[1]

The small amount, comparatively speaking, of mortality
from other causes than disease, is demonstrated by the fact,
that the annual rate of mortality to effective strength, among
European corps in India, from 1838 to 1856, was, including
all causes, according to Sir A. Tulloch—

65·6 per 1000 of strength.[2]

Dr. Ewart[3] demonstrates that the European army in India
has hitherto *disappeared*—

In Bengal, in about every 10½ years!
In Bombay, in about every 13¼ years!
In Madras, in about every 17 years!
In all India, in about every 13½ years!

To judge of this mortality, it should be contrasted with
other existing rates, as given by Dr. Guy,[4] and quoted by
Dr. Chevas.

Death-rate per 10,000 per annum at the Soldiers' Ages.

London Fire Brigade	70
Metropolitan Police	76
England, healthy districts	77
Agricultural labourers	80
Out-door trades in towns	85
Navy home stations	88
City Police	89
England generally	92
Household Cavalry	110
Twenty-four large towns	119
Manchester	124
Dragoon Guards	133
Infantry of the Line	187
Foot Guards	204
West Indies	625
India	633

[1] Chevas, 'Ind. An. Med. Sci.,' vol. x, p. 633.
[2] 'Report of the Army Sanitary Commission,' p. 12.
[3] 'Vital Statistics of the Anglo-Indian Armies.'
[4] Guy, 'On the Mortality of the British Army.'

In addition to this great *loss of life* by disease, there is the immense diminution of effective strength, occasioned by the number of men invalided and in hospital during each year.

Thus, Colonel Sykes' tables show that, between the years 1852 and 1844, the strength of the European troops of the three Presidencies was reduced in the following proportions:

Average ratio per 1000 of number invalided to strength.		*Mean of the three Presidencies.*
Bengal	36·0	
Bombay	31·60	29·4.
Madras	20·76	

Mr. Hugh Machpherson gives the annual average rate of invaliding in the European army for eight years at 26·3 per 1000, being a near approximation to that arrived at by the former authority.

The diseases which caused this amount of invaliding and sickness are, according to Mr. Machpherson:

Debility and broken-down constitution .	146	per 1000.
Rheumatic affections	142	,,
Bowel complaint	93·5	,,
Diseases of the eyes	88·7	,,
Hepatic affections	74·0	,,
Diseases of the respiratory organs . .	65·3	,,
Wounds and injuries	55·1	,,
Mental disease	35·5	,,
Fevers	35·0	,,
Heart affections	34·6	,,
Diseased spleen, and other visceral derangements	27·3	,,
Venereal affections	23·4	,,
Paralysis and apoplexy	20·9	,,
Disease of bones	19·9	,,
Epilepsy	19·5	,,
Hernia	16·5	,,
Disease of urinary organs . . .	14·1	,,
Scrofula, scurvy, &c.	14·1	,,

Although febrile affections do not appear to hold a prominent place in the above list, it cannot be doubted that by far the great majority of the individuals who die from, or who are invalided for such causes as "debility and broken-down constitution," "rheumatic affections," "hepatic affections," "disease of the spleen," &c., *would date the origin of their deteriorated health to attacks of malarious fever.* The wards of any Indian hospital, or the records of any European regiment, would sufficiently demonstrate this assertion. In fact, Dr. Ewart[1] gives the ratio per 1000 of admissions for malarious fever alone, among the men of the Anglo-Indian army of Bengal, as 1002·5, the mean of the three presidencies being somewhat less per 1000 of strength.

As it cannot be questioned that sanitary measures will, *in a great measure, prevent the occurrence of malarious disease, the conviction is forced upon us, that the sickness, mortality, and invaliding, which have occurred in India, are, at the least, mitigable* TO A VERY GREAT EXTENT.

Taking the loss by death at 65·66 per 1000, and the loss from invaliding at 29·4, the total loss per 1000 annually is 65·66 + 29·4, or 95·06!

And since each man costs the state £120 before location in India, the loss by death and invaliding per 1000 per year, reduced to a money value, is £120 × 95·06, or £11,407 4s.

Add the cost of passage home for 29·4 invalids, at £10 each, and the total loss is increased to £11,701 4s.

But, as nearly as I can calculate—and I am far *within* bounds—each man is in hospital, on the average of his service, six weeks during the year, therefore the state loses $\frac{6}{52}$ of his nominal service, which comes to the same thing as if 6 men out of every 52 (or 115·4 per 1000) spend the whole of their time in hospital, the others being supposed to be always well. The cost of treating these men in hospital for a year, at 2s. per day (and, on an average, each man will take that amount of quinine, medicine, wine, or necessaries), is £4,212 2s.

---

[1] Ewart, 'Vital Statistics of the Anglo-Indian Armies,' table xii.

Moreover, on account of these 115·4 men supposed to be always in hospital, the state expends daily pay (less by hospital stoppages), 10*d.* per day each—a sum amounting in a year to £1755 10*d.*; and even assuming that the men who die, and those who are invalided, are taken out of the class, there will still remain 20·34 men per 1000 in this class. The loss to the state on account of these men, above hospital expenses, and for food and pay, is £120 × 20·34 = £2440 16*s.*

We have then, on account of 1000 men—

	£	s.	d.
Loss of service of men who die and are invalided .	11,701	4	0
„ „ other men who remain in hospital .	2,440	16	0
Daily pay and food to men who, being in hospital, are ineffective	1,755	0	10
Actual expenses of medicine and medical comforts for keeping men in hospital	4,212	2	0
Total	£20,109	2	10

Whence, the total loss in India, which contains about 84,000 European soldiers, may be estimated at £20,109 2*s.* 10*d.* × 84 = £1,689,167 18*s.*

This, however, is a low estimate for the whole of India; the same applied to the Bengal army, where the greatest number of men are stationed, and from which the greatest mortality and invaliding occurs, would increase the total loss very considerably. The calculation, applied to each of the three armies separately, shows *that two millions of money are thus expended, or more than one fifteenth part of the whole revenue of India !* If the cost of the medical establishments were included, the sum would be doubly enormous; but that would be foreign to the purpose, as also would be the including of expenditure on account of *sick women and children.* It is merely desired to demonstrate that a large amount of expense occurs from what must be believed preventable or mitigable disease.

Still, however, this disease, and therefore expense, is not confined to soldiers alone. The following exhibits the mortality amongst European women in India, the death-rate between the same ages in England, viz., from 15 to 45, being, according to the Registrar-General's Report for 1854, 11·96 per 1000 :

Madras	.	.	.	.	24·7 per 1000.		*Mean.*
Bombay	.	.	.	.	30·5 ,,		35·477.
Bengal .	.	.	.	.	44·4 ,,		

As Dr. Chevas observes, some of these woman are Indo-Britons or Eurasians ; and it is unfortunate that, as I believe, no separate tables exist, showing the mortality of the pure European woman or soldier's wife *throughout* India. When such tables are compiled, a much higher rate of mortality may be expected to be shown, probably, as is the case in temperate climates, *nearly approaching the death-rates of the male sex.*

It is a well-established conclusion of sanitary science that infantile death-rates are among the severest and most accurate tests of the health of a community, because infants are not only the most delicate instruments for computing the force of agencies applied to all ; but many complex causes act on adult life which are not applicable to the periods of infancy. As Mr. Simon[1] observed, " it cannot be too distinctly recognised, that ' a high local mortality of children must always necessarily denote a high local prevalence of those causes which determine a degeneration of race.' "

That this is the case in India there can be no doubt. The mortality of children, up to fifteen years of age, being in the Madras Presidency, 39·8; in Bombay, 70·7; in Bengal, 84·2 per 1000 annually : giving a mean for the whole of India of 64·322 per 1000; the English rates between the same ages in twenty-four large towns being 22·36 !

According to Neison and Machpherson, the mortality amongst *officers* serving with regiments in India (Bengal), is

---

[1] Simon's Preface to Greenhow's Papers on ' Sanit. Cond. of England.'

24·032 per 1000 annually. That of officers serving at home, as with the Dragoon Guards, Household Troops, &c., being only 9·57 per 1000, and in regiments of the line, 11·000 per 1000.

According to the same authority the rate of mortality among civilians in Bengal, at the same ages—that is, below forty-five—is 17·83 per 1000.

From such statements, it appears unquestionable that the mortality amongst Europeans in India *is even now very considerable amongst the classes who enjoy greater facilities for retaining health than the soldier does,* who are better able to avail themselves of change of climate to Europe or elsewhere, and who live, as a rule, with more regard to sanitary requirements.

The chief causes of this mortality are to be found in the climate, which, from its heat and malaria, causes a constant degeneration in the European system, commencing from the period the white man first enters the tropics.

That this malarious cachexia or leucocythenic condition exerts an overpowering influence on any future attack of disease is sufficiently well known, and, probably, every Indian surgeon will have recognised that blending together of dysentery, scurvy, typhus, syphilitic and malarious cachexia, noticed by Dr. Tholozan, in the hospitals of the Crimea.

The fact that surgical patients recover with difficulty in places where marsh miasm is at all rife, that ague very frequently recurs from local injury when the patient has previously had an attack of it, and the prevalence of intemperance as a source of disease, must also be taken into consideration, as accounting for the sickness and mortality.

Thus heat, malaria, intemperance, syphilis, and the scorbutic diathesis, are the chief excitants of disease in the tropics ; *and as all are preventible, it is not too much to hope that a system of sanitation will eventually prevent the majority of affections which arise from these causes.*

Let no person imagine length of residence will climatise and prevent disease; it is the reverse. (See Chapter XXVII, " On Colonisation in India.")

To conclude, General Peel states[1]—"As a proof of the effects of climate on the soldier, I may refer to that lamentable account, recently published, which informs us that in India the 51st Regiment, in the course of fifteen days, lost one fifth of its whole strength, including its commanding officer; and that one wing of the 9th, in the same time, lost nearly one fourth of its whole number. The general commanding did everything to cheer the spirits of the men. He ordered the bands to play, but, alas! the bandsmen were numbered among the dead!"

This occurred only last year from " cholera." It cannot be too strongly impressed on those having authority in tropical climates, that measures taken for the prevention of disease will save infinitely more lives than any means which can be adopted after disease is once present. (See Chapter XVII, " On Cholera.")

---

[1] 'Home News,' Oct. 18th, 1861.

CHAPTER VI.

Advances of Medicine—Nature of the Medical Art; Homœopathy; Hygiene; and Sanitary Art: powers of the four—Sir Bartle Frere—Indian Plague.

ALTHOUGH it is undeniable that the science of medicine has made rapid advances during the last half century, still these advances have been, in a great measure, *confined to the art of diagnosis.* The means of cure which the physician of the present day is enabled to call to his aid are scarcely, with the exception of quinine, larger than his predecessors enjoyed at the commencement of 1800. The greater apparent success of the former is, however, to be explained by the fact that he places more confidence in the *vis medicatrix naturæ;* and instead of, as Liston used to describe physicians' practice, "hitting here and there, like a blind man interfering with a combat," he declines "physicking" his patient, without having a good reason for the administration of any particular medicine.

Moreover, the practitioner of the present day abjures the spoliative system of former years, and no longer abstracts blood by the pound, or "throws in" calomel by the ounce. *On the contrary, disease, especially tropical disease, affecting Europeans, is now recognised as decidedly tending to the production of cachexia and debility.* Hence, as a rule, lowering measures of treatment are eschewed.

In the 'Indian Annals of Medical Science,' No 14, is a most able article from the pen of Dr. Ewart, entitled, " A Review of the Treatment of Tropical Diseases," which clearly demonstrates that curative remedies are exhibited on a more rational system than heretofore.

As Dr. Forbes[1] observes, " There is hardly anything which is so little understood as the real nature of the medical art, and its actual power in ministering to the relief and cure of diseases. Respecting this latter point—its power—the ignorance of the lay public is literally extreme. The belief commonly entertained is that, in the vast majority of cases of disease in which the patient is restored to health, the principal, if not the sole agent in this restoration is the artificial treatment, that is, the drugs and other remedies, prescribed by the medical attendant. By such persons nature, or, in other words, the inherent powers of the animal economy, are either entirely ignored as having any share in the result, or their share in it is regarded as extremely slight and unimportant."

Hence the medical practitioner is frequently unjustly blamed in consequence of the unfortunate termination of a disease, or otherwise is lauded with that credit which should rightly be given to the *vis medicatrix naturæ*.

It is the curative powers of nature which have given so much importance to the absurd homœopathic system of medicine. This is forcibly referred to by Sir Benjamin Brodie, who, in his communication to 'Fraser's Magazine,' September, 1861, states, " If any one were to engage in practice, giving his patient nothing but a little distilled water, and enjoining a careful diet, and a prudent mode of life otherwise, a certain number of his patients would perish from the want of further help, but more would recover ; and homœopathic globules are, I doubt not, as good as distilled water."

By the previous remarks I do not intend in the slightest

[1] 'Nature and Art in the Cure of Disease.'

degree to insinuate that the curative powers of medicine are slight. The efficacy of treatment is proved in every day's practice. Attacks of acute disease are frequently cut short and subdued by appropriate remedies. Every one admits the power of quinine and other periodics over malarious fever; the influence of various remedies in scorbutic affections; the power of iron compounds in anæmic disorders; with many other triumphs of medicine, which need not be enumerated here.

I do, however, mean to submit that there are many diseases over which *curative medicine has but little power, but which hygiene is able to prevent or remove;* and the great majority of these diseases are those of the tropics.

The period has now passed when " physicians only boast the healing art." In addition to that art, they now declare that they have the means of preventing or mitigating that disease which, when established, they can scarcely cure. " Preventive medicine is the highest boon, which the medical profession has in late times offered to the state."

Among the *opprobrium medicorum* of curative medicine cholera holds a primary place. Experience, however, proves that this disease has always been most rife and fatal in those localities where sanitary matters were least attended to, and the same may be stated of typhus fever, typhoid fever, continued fever, erysipelas, hospital gangrene, gaol fever, plague, smallpox, and all diseases of a similar nature.

The want of due attention to sanitary matters in the proper materials of diet, in the absence of solar light, is equally productive of scurvy, and the numerous class of diseases, as dysentery, diarrhœa, albuminuria, and beri-beri, which attack the scorbutic patient. Attention to the same matters of sanitation will as certainly prevent and remove such diseased conditions.

Overcrowding in tropical climates will as certainly tend to induce heat-asphyxia, as the opposite condition retards the establishment of that disease. It has been frequently noticed that more attacks of *coup de soliel,* or rather heat-asphyxia, take place during the hours of sleep, when the men are in *crowded barracks,* than at other periods of *direct* exposure.

The neglect of drainage will as surely give rise to the generation of malaria and malarious fevers, as attention to this point will prevent their occurrence.

When it is considered that by far the greater number of deaths which take place in India have their origin either directly or indirectly from malarious fever, no care can be too great to obtain sites for cantonments, &c., free from all causes' calculated to generate malaria, and to carry out those principles of sanitation, under all circumstances, whether in the field, in garrison, or on the march, which will render the human system less affected by the presence of malaria. It may be truthfully stated that no European ever resides in India without becoming more or less deteriorated by malarious atmosphere.

Sufficient, perhaps, has been stated to render it apparent that, although curative medicine is frequently at a loss to check the ravages of disease, preventive medicine will *always mitigate, oftentimes prevent,* the origin of such disease. It has, however, too frequently occurred in India, as in other countries, that

> " Advice is sporting while infection breeds."

Should sanitary science become a portion of the studies of the military officer, this would quickly cease to be the case.

That the work of sanitation has been begun in India is, however, evident from the report of Dr. Grierson[1], on cholera in Scinde, in 1861, where, after detailing the causes (consisting of "impure air from the refuse and waste of organized life, combined with famine and atmospheric dryness ") the author states :—" In the great stations where the

---

[1] 'Bombay Med. and Phy. Soc. Trans.,' 7th Dec., 1861.

disease was less severe, the conditions of life were different. The regularly paid, well fed troops suffered but very little; the other inhabitants comparatively little; for in these stations, and others to some extent, *hygiene* has not been unknown. It is due to Sir Bartle Frere, K.C.B., to state, that in Kurrachee first, and afterwards elsewhere, he originated changes and improvements in this direction which only require extension, improvement, and diligent supervision, to give Seinde a character of excellence in this respect, and to protect in a great measure its inhabitants from the ill effects of epidemic states of the air. Not simply the removal of refuse, offal, waste, filth of all descriptions, is insisted on; but the disposal of all noxious matters, by inhumation, and incrimation in furnaces, constitutes the peculiar features of the hygienic measures adopted in the province."

A striking instance of the good results arising from simple sanitary measures, is to be found in the history of the mahammurree or Indian plague, which formerly ravaged the Himmalayahs, from the Snowy Range downwards.

This disease, of which mention has been made in Chap. IV, first appeared in the provinces of Kutch and Goojerat, in 1815. In 1836, it broke out with renewed force in Rajpootana, especially in the city of Pali, from whence it acquired the name of the Pali plague. Having traversed the greater part of Marwar, it became epidemic in Ghurwal and Kumaon, and other hill ranges of the Lower Himmalayahs, whence, in 1853, it finally disappeared, under the influence of sanitary measures which the affected villages were bound to adopt.

CHAPTER VII.

In addition to the expense referred to in Chap. V, there
is another source of immense loss to the state, in endeavours
to improve naturally unhealthy localities, which were fixed
upon in former years, either from ignorance of the sanitary
requirements of a station or from political necessity.

Not a few of such stations have been mercifully abandoned
by order of government, as cantonments for European
troops, and as instances, Berhampore, Kurnaul, Loodiana,
Lahore, Masulipatam, Secunderabad, Bhooj, and Kaira may
be mentioned.

The descriptions of these stations show that they are sur-
sounded by low irrigated land, marshes, or jheels. Kurnaul,
for example, is close to a canal which is higher than the
surrounding country, and hence no drainage can take place;
Loodiana is on the banks of a dry water-course, formerly the
channel of the Sutledge; Lahore or Anarkullie has on one

side a stagnant nullah, and on the other Mahometan grave-
yards, and also lies lower than the surrounding country;
Masulipatam is situated at the mouth of a branch of the
Krishna, and surrounding the fort is a saline swamp; Kaira
is low, surrounded by irrigation, and the country traversed
by water-courses; Bhooj is surrounded by the 480,000 square
miles of the Runn of Kutch.

Such concomitants would, it is hoped, at the present period
prove insurmountable objections to the establishment of
European troops in the locality.

According to Mr. Wells, previous to the year 1846, the
death-rate in Bombay never fell lower than 101·0 per 1000.
During 1812, it was 203·7 per 1000; and the average of
eight years was 147·2 in the 1000 of European troops
stationed there.

In 1845, in consequence of this excessive mortality, the
majority of the forces were located at Poona, and Kirkee.
Dr. Coles[1] shows that during the years 1855-6, the death-
rate was reduced at those stations to 9·0 per 1000 !!

Poona, it may be observed, is the head-quarters of the
army during a considerable portion of the year, is situated
in the comparatively speaking salubrious table-land of the
Deccan, and has some pretensions to a satisfactory sanitary
condition.

We have suffered, and indeed in many instances are now
paying the penalty of faults committed years gone by, in
choosing unhealthy localities for cantonments, stations, and
public buildings.

The following references to government orders (the
majority of recent date) sufficiently prove that the principles
reiterated over and over again by army surgeons—"that it
is cheaper to keep veterans in health and readiness for service
than to be ever recruiting and training new men, and that

[1] "Hospital Statistics of the Bombay Presidency;" 'Bombay Med. Phy.
Soc. Trans.,' 1857.

there is an irreparable condition between the preservation of efficiency of armies and submission to the laws of medical and sanitary science "—are now thoroughly acknowledged.

'Public Works Code,' chap. v, sec. 1, par. 3, states :—"The health and comfort of the troops are to be held as paramount considerations, to which all others must give way."

Again, 'Government General Orders,' Aug. 8th, 1853, states :—" It being essential to the health of the troops that great attention should be paid to the position and aspect of all barracks and hospitals, the superintending surgeon of the division, or, in his absence, the senior medical officer of the station, shall, previous to the laying of the foundation of such buildings in all future cases, be invariably consulted on the subject, and commanding officers will conform to the opinion of such medical officers given in writing, or refer the question should they see cause, for the decision of government."

It is also ordered in the 'Code of Regulations of the Public Works,' that committees, composed, when practicable, of two experienced medical officers, in conjunction with a military officer, must be assembled to select sites for barracks, hospitals, &c.

The 'Code of Regulations for Public Works Department,' 1858, chap. viii, p. 82, enforces on the executive engineer that he must examine the position of neighbouring buildings, cultivation, tanks, jheels, jungles, note the direction of prevailing winds, and the character of the country over which they blow, and must obtain the opinion of medical officers as to the salubrity of the site, with reference to these and all other circumstances.

The 'Medical Regulations for the Army of 1859,' par. 9, page 79, states :—" Before any new barrack or hospital is erected, the plans and site will be submitted for approval, in so far as regards the healthiness of the building, to the Director-General."

Separate general orders, under the head " Cantonments

5

and Quarters," provides for the sanitary condition of the former.

And so on might be quoted orders on almost every subject connected with sanitation, the *spirit* of which should be invariably carried out.

Lest it should be objected by a senior and superior officer, that a junior and subordinate officer is not qualified to offer advice on sanitary matters, and lest the medical officer should be again requested to confine his attention to his hospital and patients therein, the army medical school has been established, in which future army surgeons—after their reception by the usual examination into the ranks of the medical profession—will be instructed in all causes which specially affect the health of the soldier. There army surgeons will learn *all that is known* respecting the preservation of the soldier's health. Diet, clothing, education, habits, peculiar duties in times of peace, situation and circumstances in time of war, principles of ventilation, warming, draining, cleansing, architectural arrangement, influence of geological formation, epidemics and their causes, camp diseases, form only a small part of the special knowledge, now to be conveyed to the young medical officer; in fact, he is fortified at his entrance into the service with that special knowledge which others before him have only acquired after years of experience and research, and is rendered capable, at the onset of his career, of advising on all sanitary subjects.

The truth of the heading of this chapter requires, I think, no further demonstration.

CHAPTER VIII.

THE SANITARY CONDITION OF INDIAN STATIONS.

In this chapter I shall quote from a few published reports,
or from communications I have been favoured with from
various medical officers; and as the importance of hill climates
as a means of preserving the health of Europeans is becoming
daily more evident, I shall first refer to a few of these
localities.

Dr. Mackay[1] reports, the station of Ootocamund, in the
Neilgherry mountains, is divided by a deep valley, a large
portion of which is filled with water, forming a lake. "At
the upper extremity of this, the large station bazaar is situated;
the houses there are crowded together, those in the lower
street are built close to the water, many of them on a foundation formed by the rubbish thrown out from the houses above,
and the filth washed down by the rains from the upper street.
With every natural facility for doing it effectually, drainage
has been totally neglected, not only in the bazaar, but
throughout the whole station. It will hardly be believed that
undrained swamps still intersect this important sanitarium.
The invalid may shoot snipe in some of these without leaving
the compound of his house. Some of the best houses are

[1] 'Madras Journal of Medical Science,' vol. v, p. 61.

built over these swamps, some on a level with them, and not
a few on a lower level than the swamp itself." The potato
cultivation, the noxious weeds, nettles, brambles, giant
thistles, and lobelias, are described as most luxuriant and
objectionable—" In short, the whole station is a strange
mixture of snipe-bogs, neglected compounds, and neat flower-
gardens. Every convenient bush is made use of to deposit
filth under, of every description. The sweepings and refuse
of each dwelling are thrown wherever they can be conveniently
disposed of. It is argued that the station has been healthy,
that such things are better left as they are, that no injurious
consequences can result in this region." Dr. Mackay states,
however, that the deaths are twice as numerous in the lower
as in the upper street of the bazaar. Also, although he has
only seen one case of cholera on the hill, "yet, should the
preparation for its reception continue, in the way of a disregard of all sanitary arrangements, there is every reason to
fear that cholera will some day exhibit its virulence on the
Neilgherries as it has done in other temperate climates."

Prophetic words ! but, like Cassandra's voice, unheeded ! !
The ' Neilgherry Star,' of November 13th, 1861, reports,
" Another case of cholera terminated fatally in the dispensary
last Wednesday. The dreadful disease has now unmis-
takeably made its appearance among us."

According to Mr. Grant,[1] " Nothing could be worse than
the state of conservancy at Simla." He describes the smells
along all the bye-paths as being most disgusting, from accu-
mulations of human ordure, offal, and dead animals in the
ravines that intersect the station ; in fact, a native population, amounting to 10,000 or 12,000 people, were scarcely
restrained by any police rules, and hence an amount of
nuisance scarcely credible.

That the sanitary condition of Simla is not much improved
in the year of grace 1862 is evident from the subjoined in-

[1] 'Ind. An. Med. Sci.,' 1852.

formation afforded me by Assistant-Surgeon J. J. Clarke, now in medical charge.

Dr. Clarke writes (10th January, 1862) :—"The native bazaar is built of tiers of houses on the south side of the hill, and cuts the station of Simla in two, with about the same number of European houses on the eastern as on the western side of it. Its main street, which is a thoroughfare, and the high road from Chota Simla to Boileau Gung, is kept clean, and is drained by means of open drains, which flank it on either side. Its lower streets are, for the most part, filthy, badly drained, and with no well-adapted means for clearage. It is densely populated, and the nuisances of this large population are allowed to ooze and dribble forth of themselves, finding their way into the several 'nullahs' which run down the side of the hill; so that, in truth, the whole filth of this extensive bazaar lies hoarded in these several cesspools during the hot season until the rains, impregnating the air with emanations most detrimental to health.

" There is no conservancy officer here; and neither the local nor municipal authorities appear to have instituted any measures for providing for a complete system of drainage."

When this is the condition of our principal Himmalayan hill sanitarium—only so in name—no wonder that diarrhœa is there so common, as to have obtained the name of " Simla trots," that Dr. Mackinnon, Deputy-Inspector-General of Hospitals, should remark—" Of late years Simla has rather retrograded in public opinion as a sanitarium."[1]

Of Nynee Tal it is stated :—" The Aya Puetun side of the lake is dreadfully damp; hundreds of trees might be felled with advantage to those who live on that side; a place so shut in by nature as is Nynee Tal requires no superabundance of trees, and still less of jungle, to make it unhealthy; and those who have passed a rainy season there, can hardly

[1] 'Report on the Extent and Nature of Sanitary Establishments for European Troops,' p. 73.

forget the one drawback to a residence in that beautiful spot—the cold damp smell that arises before sunset, and continues throughout the night, which a dozen sharp hatchets might remove in a short time.[1]

A late writer[2] states of the same station—" Let the brushwood and jungle be cleared away, leaving only so much timber as is necessary for fire-wood and for scenery. If any one doubt how unhealthy Nynee Tal is becoming, let him go into the grave-yard and count the number of new graves, and then let him go into the soldier's hospital, and see the pale yellow faces of men in whom hope has died away, who are worse than when they came up. . . . Another great nuisance of Nynee Tal I have alluded to, I mean the rank state of the weeds in the lake, and the jungle round the lake. To walk round the lake on the lower mall is quite enough to give a weak man a dangerous fever, so great is the malaria."

Dr. Porter writing to me from Nynee Tal, January 25th, 1862, gives an account of the conservancy arrangements at this station, which consist of the objectionable method of cess-pools *emptied every two or three days.* The drainage is stated to be excellent, from natural configuration of the ground. Dr. Porter, however, adds :—" I regret to say that the jungle is very thick in the civil station, but the cantonment is properly cleaned."

" The Hills," describes Darjeeling as " a dismal hole, subject to fever on the spot." Smallpox is stated to be there an indigenous disease, and is described by Dr. Collins, as " the scourge of these hills." It may be recollected that Miss Nightingale, at page 17 of her 'Notes on Nursing,' while dissenting from the doctrine that disease, having once originated, is handed down from one generation to another, declares that she has seen and smelt variola, forming in

---

[1] " Kumaon and its Hill Stations ;" ' Calcutta Review,' June, 1856.
[2] ' Times of India,' Sept. 7th, 1861.

various localities, the abodes of filth, and destitute of sanitary arrangements.

In my 'Annual Report of the Mount Aboo Sanitarium,' dated 1st January, 1862, after depicting the beauty of the station, the fact of its height being sufficient to carry it above the range of the hot winds of the plains, the temperate range of the thermometer, the excellent condition of the water, the vegetables which the climate permits to flourish, the perfect freedom from cholera which the mountain affords, the small amount of diarrhœa, and the mild type of the malarious fever occurring, I proceed as follows :

"Thus, therefore, it may with truth be stated, that Mount Aboo presents most of the essentials for a sanitary station. I regret, however, being obliged to repeat the statement made in my last 'Annual Report,' that the labour of man has not fully developed and increased the advantages and capabilities of the locality, and that sanitary science and medical hygiene are not yet made available to the extent which I again respectfully submit ought to have been the case."

It would scarcely appear credible that numerous individuals, probably from ignorance, view with contempt and neglect the most common sanitary precautions. Such persons by their good fortune, or, perhaps, from a naturally powerful constitution, having escaped disease for a considerable period, almost forget that they are mortal, until, sooner or later, having imbibed a large dose of the poisoned air, they call upon the medical man to cure them of that disease, which prior attention to his *ex necessitate* superior sanitary knowledge would have, in all human probability, utterly prevented. To such persons, I would say, study the different sciences on which the preservation of health in tropical climates depends, the etiology of disease, physiology, chemistry, geology, &c., &c., and *then*—act as you think fit. As Dr. Johnstone truly observed, "A blockhead can deny more in a single hour than one hundred doctors have proved in one hundred years."

As Dr. Mackay writes of Ootocamund, I have also heard it argued with regard to Mount Aboo, viz., that it is better to let such things alone; that in such a climate no harm can result; that as matters have always been the same, it is better to allow them to remain *in statuo quo—"Ignorantia facti excursat."*

When the feelings of a community, with regard to preservation of health, are represented by such expressions and writing as—" How do you think we managed to exist and enjoy good health before all this *fuss* about sanitary reform?" "Wait till the rains and all the dirt will be washed away!" "Why should I clean my compound when only here for a short time?" When a gentleman, just previous to the rains, makes use of the abandoned commencement of a well to deposit *débris* in, and when this is declared no nuisance! When this represents the feelings of a community, how can anything excepting *orders* from superior authority be expected to produce any very beneficial result.

Notwithstanding improvements effected during the brief existence of an honorary sanitary committee, Mount Aboo now remains with only imperfect efforts at drainage; deciduous vegetation arises and decays almost unchecked; the water from the lake is allowed to overflow large surfaces, rendering them malarious in the extreme; there are no public latrines; bushes and ravines are made into native temples of Cloacina; and there is not, nor ever has been, a single public scavenger entertained for the civil portion of the station !

Such being the case, as I stated in a report on the subject, "Mount Aboo is dirtier now than it was when first resorted to by Europeans, thirty years ago," or when first described and recommended as a sanitarium by the late Sir Alexander Burnes.

Any person conversant with the laws of disease in tropical, or, in fact, in any climates, will not be surprised to learn that malarious fevers are rife on Mount Aboo to the extent, during the period from 1855 to 1860 inclusive, of 1222·74 per 1000 of strength annually ! The admissions to strength throughout the armies of Bengal, Bombay, and Madras, being respectively 1002·5, 631·9, 284·6.[1]

While, however, the admissions are so greatly in excess, the mild type of the fever is forcibly demonstrated in the fact that, for the same period, the deaths at Aboo from malarious fever were only 1·55 per 1000, while in Bengal they amount to 10·08; in Bombay, to 4·9; and in Madras, to 1·6 per 1000. *Malarious fever presenting so mild a type may certainly be easily removed by sanitary measures.*

In considering the amount of fever prevalent amongst the men on Mount Aboo, it must be recollected that although many of the soldiers were convalescent, sent up for the recovery of health, and thus, therefore, more liable to relapsing attacks, still a certain number of robust and healthy duty men are included in the returns.

When I state that the prevalence of malarious disease among the men of a detachment of her Majesty's 33rd Regiment, not composed of convalescents despatched to Aboo for the restoration of their health, but of effective soldiers stationed here on garrison duty, was, during November and December, 1860, and January, 1861, respectively 194·0, 373·1, and 328·3 per 1000 of strength, and during May, September, and October, 1860, 50·0, 50·0, and 88·2, there will, I think, be little doubt that sanitary art is required on Mount Aboo.

If matters are allowed to continue as they are, future medical officers will have to report not a mild, but severe type of malarious fever; not isolated cases of typhoid, but epidemics of that disease ; not immunity from cholera, but its ravages, on the excrete-loaded ground of the mountain station.

---

[1] Ewart, 'Vital Statistics of the Indian Armies,' table xii.

The same want of public scavengers would appear to exist at Mahableswar. Dr. Lord, the latterly appointed super-intendent of that station, informs me that "the cleansing of that station is a matter more of private than public arrange-ment. The avenues and gardens are all kept swept by the 'mollees,' and the usual domestic arrangements provide for the removal of the filth from the immediate vicinity of the dwelling-houses, as well as from the hospital and sanitarium." Those conversant with Indian "domestic conservancy," will be at no loss to understand what becomes of the *débris* and ordure. In the absence of public bazaar latrines and public conservancy establishment, it becomes mixed and incorporated with the surface-soil, and forms the ground, which sometime or other cholera and typhoid may infect.

Inspector-General Mackinnon[1] reports of Landour:—"The barracks are ill adapted for the purpose for which they are in-tended. They are all bungalows, and are used to accommodate from sixteen to twenty men or more each. All are narrow, low-roofed, and without any ventilation, except what is ob-tained through the doors and windows. When convalescents are crowded together, as they often have been in these bunga-lows, they have not a fair chance of regaining their health."

At Subathoo[2] the barracks have been erected for several years, and have all the faults in ventilation and construction to be met with in our older barracks.

Mr. Bradshaw, writing to me from Dugshai, January 9th, 1862, thus describes the hospital at that station:

"The hospital for men has also an excellent site, but is built, in my opinion, upon a very faulty plan. It is like a barn, is divided into large rooms, ventilated and lighted by rather small dormer windows; and between these rooms and the verandah are long, narrow, gallery-like rooms, which appear to shut out, to a great extent, light and air from the

---

[1] 'Report on Extent and Nature of Sanitary Establishments.'
[2] Op. cit.

interior of the building, rendering it in rainy or cloudy
weather very gloomy, cheerless, and close."

For a description of the married men's quarters at this
sanitarium, I refer to Chapter XXVIII, " On the Soldier's
Wife, Widow, and Children."

The barracks at Mount Aboo are built in a valley, overhung
by rocks 1000 feet high. The valley communicates with the
lake by a ravine, and is insufficiently drained. In addition
to this, the buildings are placed gable-end on to the prevailing breeze, *and in a direct windward position to No.* 1 *both
wash-house, urinal, and latrines are placed!*

After this detail of sanitary defects, it affords me pleasure
to be able to quote accounts of other hill stations, which
have pretensions to be called sanitaria.

Dr. Sinclair, in a letter to me from Kussowlie, states of
that station :—

" Drainage is chiefly natural, and where required, temporary drains have been dug and sewers made.

" The south side of the hill is but sparingly wooded, and
around the bungalows the trees have been well cleared.

" The conservancy arrangements are good. They consist
in constantly cutting down the rank vegetation, in lopping
off the lower branches of the trees to permit free circulation
of air, the strictest attention to cleanliness in and around
the barracks, and removal of all filth to a long distance, preventing the natives committing nuisances, &c. Latrines are
constructed for the use of the inhabitants of the regimental
bazaar, and occupants of houses are also recommended to have
the same for their servants. The state of the bazaar is also
attended to, no filth nor pools of water being allowed to accumulate.

The grave sanitary defect existing at Mahableswar, in the
absence of public scavengers, has been previously referred to.
In other respects this hill station appears to be well looked
after. Dr. Lord, in the communication previously men-

tioned (dated 8th March, 1862), " The compounds here are
generally large, varying in extent from fifty to five acres, and
between them large tracts of untenanted ground.   Houses
are, therefore, far apart, generally placed on elevated sites,
selected as points of view, &c., and the buildings are mostly
placed towards the west, as from that direction we get the
soft and cool sea breeze.   The easterly or land winds are
harsh and dry at some times, and slightly heated at others.
Sometimes they are also cold and cutting, so that people
avoid exposure to them in building houses.   The sea is visible
about fifty miles distant from many of the residences.   The
bazaar is the only collection of houses, and the cleansing of
this is under the ' Foujdar.'   It is only a small place, and
four sweepers (kept up and paid from local means) are
generally found sufficient for keeping the streets clean.

" In the days when the Chinese and Malay convicts were
kept in the gaol here, the roads were swept and kept as clean
as the avenues to the houses, but now we have no such means
at our disposal, and must content ourselves, after the leaves
have fallen (about the middle of April), with one general
sweeping of principal roads, which must be done by hired
labourers, paid for out of the road fund, which is liberally
supported by voluntary contributions from residents, house-
owners, and visitors at the hills.

" No malaria is believed to be generated here, and there is
no clearing of the jungle encouraged.   Water is abundant,
and from the hill by the village of Mahableswar no less than
seven rivers—the Krishna, and six of lesser note—are said
to take their rise."

Dr. Leslie informs me that " the military station of Wellington (formerly called Jackatalla) is two miles distant from
the civil station of Coonoor, in the Neilgherries.   It is situated
in a large amphitheatre of hills, which opens towards Coonoor.
Magnificent barracks are erected here to accommodate 1200
men, with every necessary appurtenance.   The supply of

water is abundant, and of pure quality. The drawbacks to the locality are, that the heat is sometimes great, and the winds occasionally very harsh.

" Coonoor, altitude 6000 feet, is situated on the summit of the Ghaut. The residences are situated on a series of hills, intersected by wooded ravines, through which beautiful streams flow over rocky bottoms."

The climate of Coonoor is equally agreeable, more so, it is stated, than either of the other Neilgherry hill stations. Public scavengers would, however, appear to be a want here as elsewhere.

Dr. Sparrow informs me of Landour—" During the rainy season an extra conservancy establishment is entertained. In my belief, there is not a better drained or cleaner station in India. The bazaar is well kept, well drained, and clean ; this is owing, in my opinion, to the fact of its being under military, and not civil control."

Any person perusing the foregoing descriptions of the sanitary condition of hill stations, will probably be amazed that such continues to exist, but the reason has just been stated by Dr. Sparrow. Those sanitaria under military control alone are, I have reason to believe, generally cleaner, and in a better sanitary condition than either where divided authority or civil control alone exists. This arises from the fact that hill stations are made the summer residence of civil and political officers, who leave for their various duties during the cold weather, and on many occasions are absent during the rains also. Such individuals only seeing the station when all appears dry, and when the malarious season has passed, either do not believe the deleterious effects arising during other periods, or, if so believing, congratulate themselves that *they* will not be present during the unhealthy period. In the mean time, they add to the general sanitary defects, which cannot fail, sooner or later, to originate epidemic disease.

It cannot, of course, be expected that government should be at the expense of sanitary arrangements and establishments for hill stations frequented by visitors, civilians, and others, who go there for their own pleasure and amusement, or to escape the heat of the plains below. The most that the state can afford is, the means of sanitation for *purely military* hill cantonments, and *authority* for the imitation of the same at other places. The expense of the latter should be met by visitors to the hills, and by those individuals who make a profit of letting houses to such visitors. On either one or the other, or on both, a compulsory tax should be imposed, and the proceeds devoted to road-making, drainage, clear-ance, public scavengers, the erection of latrines, and all other sanitary requirements. Such an imposition would not be felt by the classes who can afford the expense of a yearly journey to a hill retreat; neither would it weigh heavily on those house-owners who make large profits by exorbitantly rented bungalows. The tax, however, should be *compulsory*, and levied by *authority*, otherwise, if optional, many would decline payment. Such a tax should not extend to military officers present on duty, or who have been sent up on sick certificate, or to occasional visitors, who may only remain on the hill for a few days. These classes might well be exempt, both on pecuniary and just grounds.

The proceeds of the tax should be placed at the disposal of a sanitary committee, or, where such cannot be formed, should be under the control of the station medical superintendent, or, when this functionary has not been appointed, should be administered by the medical officer in charge.

The same officer should also have under his superinten-dence a number of prisoners, from some of the overcrowded gaols in the plains, who, after hutting themselves, would suffice for the sanitary work of the place, a lesser number being re-tained after the station was once placed in thorough order. This plan would not entail extra expense on the state, and its beneficial results can scarcely be considered in too favorable a light.

With regard to cantonments in the plains of India, it may be stated that the great majority exhibit grave sanitary defects, either in their site or in the position and structure of the public buildings. Thus, as Dr. Chevas[1] states, " we have a set of stations where the mortality never falls in any one year, even after a pestilential visitation, to the highest English standard, and where in many years the sickness and loss of life are truly appalling." Amongst these are Dinapore, the decennial range of sickness being from 23·037 to 110·755; Cawnpore ranges from 23·933 to 145·679; Chirisurah from 26·316 to 142·857; Fort William range 39·773 to 88·710. Dum-Dum 31·153 to 200·280; Bombay and Colabah, between the years 1830—9, 109·20; the last, however, of which, owing to the removal of troops to Poona, as before referred to, and attention to sanitation, is now much healthier; and in 1850—1, according to Dr. Coles,[2] not showing a greater mortality than 79·8 per 1000.

The unhealthiness of Bombay is attributed in a great measure to the over-population of the island, and to the proximity of salt marshes in Salsette and other more immediate localities. The barracks, moreover, at Colabah are situated on a low, narrow spit of land, jutting out into the sea, which, from the lowness of the site, was only kept out by embankments. (See Chapter XIII, on Barracks.)

Dr. Lownds,[3] in his article entitled " Sanitary Aspect of Bombay," observes—" At the present time, the drainage of Bombay, both as regards ordinary house-drainage and the monsoon water, is as defective as can be. Even close to the public roads we see vile attempts at drainage which would disgrace a nation just emerging from barbarism."

[1] 'Ind. An. Med. Science,' vol. xii.

[2] " Hosp. Stat. of the Bomb. Pres." 'Bomb. Med. Phy. Soc. Trans.,' 1857.

[3] ' Bombay Quarterly Review,' 1858.

Although much improved during the last few years, the sanitary state of Bombay is most imperfect.

Calcutta, the spot on which the fortress of Fort William is built, and where the soldiers' barracks are located, was a most pestilential swamp, and although now thoroughly drained, and therefore in the cool season harmless enough, no sooner does the rainy season commence, than the excessive heat and moisture call forth unmistakeable signs of miasmata. "The soldiers' barracks are also so constructed, that it is impossible for the cool breeze to pass through them during the hot season, and they are not placed high enough to be above the reach of marsh miasm. The consequence is, that the mortality of soldiers in Fort William is great."[1]

Dum-Dum, again, is surrounded by "paddy fields," and is so ill drained, that, in the height of the rainy season, one cannot pass from one part of the cantonment dry-shod, except on made brick and "concker" roads. The barracks, however, are infinitely better than the neighbouring ones of Fort William.[2]

Dr. Chevas states that the new European cantonments at Agra occupy so dead a level, that nearly the whole of the rain-water is absorbed into the ground where it falls, and it has been found necessary to make raised foot-paths, to enable the men to pass from one barrack to another dry-shod. The consequence is, malarious fevers are rife among the troops who occupy "the beautiful and costly barracks" erected there.

Inspector-General Mackinnon reports of the new cantonment at Cawnpore, that much of the ground is so level as to render it difficult, if not impracticable, to drain it thoroughly; the water never appears to drain away, but percolates slowly through the spongy surface.

The same is said to be the case, to a greater or less extent, at Mean, Meer, Allahabad, Barrackpore, and in many other places.

---

[1] 'Report of the Army Sanitary Commissioners.'
[2] Op. cit.

In my report on the medical topography of Bhooj, published by permission of the Director-General of the Medical Department, in the 4th number[1] of 'The Bombay Medico-Physiological Society's Transactions,' after a full description of the barracks at that place, I state :—"All things considered, I am constrained to form the opinion that these barracks are unfitted for the residence of European soldiers; not, however, so much from the faults inherent in their construction, as from their unfortunate situation and locality, being, as they are, close to a heat-absorbing sandstone rock, fronting the opposite direction from which the wind blows during ten months in the year, and being still more shut out from the breeze, the one by the other, and the latter by the gun-shed. It is difficult to imagine what military reason could exist for placing these barracks in their peculiar position; and it is equally impossible to opine what hygienic or sanitary recommendations led to this spot being fixed upon for their location, particularly as, half a mile or more to the south-west, there is a dry, elevated spot, which, as far as can be the case in the Bhooj cantonment, presents all the desideratives required for a barrack site."

It would, however, be an endless and ungracious task to quote more sanitary defects of site or buildings. It must be recollected that political reasons, in former days, led to many of our stations being fixed in unhealthy localities; on the banks of movable rivers, in the neighbourhood of large cities, or on low-lying lands. Military necessity must also have caused the erection of many barracks in their present position, but some, as is the case with the buildings at Bhooj and Aboo, seem to have been constructed in downright ignorance or defiance of sanitary laws.

At the present period, with an era of peace inaugurated over India, with the knowledge of past errors, and the

---

[1] New Series.

consequent sacrifice of human life, and enormous expense thereby entailed, it behoves us to endeavour to place our troops on the most healthy sites, and in the most scientifically constructed buildings.   We may remain assured, as regards disease, of the truth of the quotation—

" Sublata causa tollitur effectus."

CHAPTER IX.

ON HILL SANITARIA.

Importance of the Subject; long since recommended by Medical Men—Establishment of the first Sanitary Stations—Use of Hill Climates—Existence of Malaria on Hill Ranges—Cholera—Dysentery—Opinions regarding the location of Troops on Hill Ranges : Objections refuted—Proposal to send healthy men to Hill Sanitaria—Peculiarities of Hill Climate : Temperature—Use of Mercury—Necessity of Sanitation in Hill Climates—Diseases of Hill Climates—Conservancy—Clearing—Drainage—Site of a Hill Station—Barracks at Hill Stations—Height of Hill Stations—Danger of the Ferae—Locality for the Supreme Government: Bombay recommended; Reasons—Lady Canning.

AMONGST all the numerous means which are required to be adopted to preserve the health of Europeans in India, there are none of such vast importance as the establishment of hill stations, or sanitaria. Now that it is evident that a large European force must be retained in the country, the subject of the location of troops in hill climates is receiving more attention than heretofore. That this is the case is evident, from the interest which the present Commander-in-Chief of India and the Commander-in-Chief of the Bombay army evince in the matter. That considerable progress has latterly been made in the establishment of sanitary stations, in various localities, is also shown in the ' Report of the Extent and Nature of the Sanitary Establishments for European Troops in the three Presidencies,' recently published by authority, where a more or less detailed account of some thirty of these stations is given. Much, however, yet remains to be effected; and I venture to prophecy that, as time develops the advantages of a hill climate, combined with a strict sanitary system; as railroads traverse the land; as civilisation advances; as the natives of India, becoming more aware of our resources, cease to imagine the *local* absence of European soldiery results from decline of power—that eventually the great majority of our soldiers will reside on the hills, and that it will come to be the exception, and not the rule for a European regiment to be stationed on the plains.

Medical men have long since urgently recommended the location of European troops serving in tropical climates should be effected on elevated regions. Thus, Dr. Lind devoted a portion of his work, published in 1768, to the selection of healthy spots, and recommended a removal into a temperature, " where the heat of the day seldom exceeds 80°, and the cold of the night is about 54°." Jackson and Hunter wrote in much the same manner.

The establishment of the first sanitary station in the Himalayas appears to have been due to the recommendation of Dr. Gilb, of the Bengal army, who, as early as 1820, was exerting himself in this direction. This officer was ably seconded by Julius Jeffreys, who, in 1824, wrote his ' Essay on the Climate of the Hill Provinces of the Himalayas.' In consequence of this essay the attention of government, was more immediately directed to the subject, and the military sanitary stations of Simla, Missouri, and Landour, shortly afterwards fixed upon.[1] Now, as before stated, this number has increased to upwards of thirty.

As Dr. Mackay[2] observes, " Both the profession and the public have shown a tendency to overrate hill climates in India, with reference to their suitability for sanitary stations. I feel convinced that the true value of these hills will ultimately be found to consist in their possessing a climate in which Europeans can *maintain* their health and vigour.

[1] Jeffreys, ' The British Army in India.'
[2] Mackay, ' Ind. An. Med. Sci.,' vol. v.

They are, therefore, better suited for European settlers, and stations for troops, than for depôts of sick."

Dr. Murray stated his opinion, that there is nothing more likely to bring the Indian hill stations into disrepute than an over-estimate of the merits of their climate—"It is calculated to convey an erroneous impression of their qualities, and to raise up expectations which must end in disappointment."

Dr. Morehead wrote :[1]—" The hill climates are not merely to be regarded as conducing to the recovery of suitably selected convalescents, but also as materially assisting in maintaining, at a high standard, the general health of European soldiers and their families." The same authority in his ' Report on the Sanitaria of the Poona Division of the Army,' dwells much on the *injurious* action of the cold season of hill climates in certain forms and stages of disease.

In another place :[2]—Dr. Morehead remarks, " The cold season of all hill climates of India will excite dysentery in cachetic individuals, irrespective of the conditions of malarious generation."

In my ' Manual of the Diseases of India,' chap. iii, ' On Hill Sanitaria,' I state :—" Although the curative properties of hill climates are thus limited, I believe that the preventive powers are almost unlimited; and hence I must enter my humble protest against the hill climates of India being *only* regarded as convalescent depôts. Broadly speaking, organically diseased patients should never be sent to the hills."

Every medical officer who has had charge of a hill station will readily admit, that *invalids only do well in such climates when no internal organ has become structurally diseased ;* that abdominal affections, as diarrhœa and dysentery, are invariably rendered more severe by removal to hill stations ;

[1] Morehead, ' Letter to the Right Hon. Sec. of State for India,' 17th Sept., 1861.

[2] ' Clinical Researches,' vol. i, p. 65.

that affections of the liver are rarely benefited; that rheumatisms are frequently increased; that the cold stage of fever is often prolonged by the greater cold of the climate; that affections of the brain are not improved; that diseases of the respiratory organs are generally aggravated; and that neuralgic complaints are more painful.

The class of cases benefited by removal to the hills are cases of debility after fever; fevers recurring at particular seasons on the plains; all cases of local cachexia; or splenic leucocythæmia, provided the spleen is not permanently large; syphilitic cachexia, if unattended with rheumatism, and all cases of debility after cholera.

It is, therefore, evident that the capabilities of the hill climate as a curative agent are but exceedingly limited.

Although the idea held by certain individuals that malaria cannot exist at an elevation of 4000 or 5000 feet above the level of the sea is a complete fallacy (see chapter on Malaria), still it is equally a fact, that the malaria generated on the summit of mountain ranges, while sufficiently powerful to induce repetition of paroxysmal fever is only in comparatively exceptional instances capable of originating intermittent disease *de novo, in individuals who have not become deteriorated and cachectic by residence on the Indian plains.* Hence, an individual coming from Europe, and being located as soon as possible in a hill climate, would escape malarious degeneration, malarious fevers, and all the secondary affections arising therefrom, simply because, in the majority of instances, the European's natural vigour of constitution would enable him to withstand the diluted malaria of the mountain-top; and that it is diluted, is certain, from the mild type of intermittent which it alone induces. (See Chapter VIII, Sanitary Condition of Mount Aboo).

Again, although cholera has occasionally visited hill stations in the Himalayas, yet there are other mountain ranges where it is unknown, as, for instance, on Mount Aboo, and, until last year, at Ootocamund. Even although the neglect of sanitary regulations may have prepared the way for cholera on some hill stations, still it cannot for one moment be thought that this disease visits these elevated regions nearly so frequently as the arid plains below. *Hence the European located on the mountain has every chance of escaping death by cholera.*

Dysentery also is not near so fatal or prevalent on mountain regions as in the plains below, a tendency to determination to thoracic organs usurping the abdominal affections consequent on the greater heat of the lower climate.

Thus, notwithstanding it appears beyond doubt that Europeans on hill climates, who go there in health, are infinitely less liable to contract disease than others who remain on the plains, yet our greatest authorities decide against the location of British troops on hill stations.

Dr. Morehead states:[1]—" To place **permanently** at such elevations as Ootocamund all the British troops in India, even if politically practicable, would not prove, in my judgment, the best method of fitting the European soldier for the maximum of efficient service with the minimum sacrifice of health and life. Doubtless, Europeans permanently residing in a hill climate, such as Ootocamund, would retain much of their native vigour, but they would not be efficient for the contingencies of military service in India. If suddenly called to the plains for service in the hot season, there would be a heavy sick list from seasoning fevers and biliary derangements, and a rapid loss of vigour and stamina would result. Then, the service over, and the men exhausted by heat, fatigue and sickness, moved back to their hill cantonments, would be subjected to much mortality and invaliding from these forms of disease, for which the cold and rainy seasons of hill climates are unsuitable. Were it possible to transport, *in a few hours,* troops from the camp at Aldershot to the

[1] ' Letter to the Secretary of State for India,' Sept. 17th, 1861.

plain of the Ganges, any time between March and November,
use them for active service, and return them, broken in
health, *in a few hours* from India to Aldershot, any time
between October and May, the result need not be told. It
would be analogous in kind to that which may be anticipated
from the permanent location of troops in such positions as Ootocamund and the interior Himalayas, and their rapid transfer
any time between March and November, in the one instance
to Malabar, the Concan, or Carnatic, and in the other to
the plains of the Punjaub or of the Ganges, with their return
to their hill cantonments between September and April."

In proof of the heavy sick list which would result from the
conveyance of troops from the hills to active service on the
plains, Dr. Chevers brings forward the cases of the 1st
Fusiliers, 2nd Fusiliers, and 75th regiments.

The 1st Fusiliers were ordered to march on May 13th,
1857, on the breaking out of the mutinies, from the Himalayan Hill Station of Dughshai (elevation 5,600 feet) to
Umballa. They started the same day and marched in shirt-
sleeves, and halted at Kalka (fourteen miles), where *cholera
was prevalent;* at 1 a.m. the following morning they marched
on Pinjore, when, after making a short halt, they proceeded
to Chundeeghur, arriving there at half-past 6 a.m. After
resting all day, they marched at 5 p.m., and halted at 10 p.m.
at Mobarackpore, where *cholera was also prevalent,* and made
its first appearance in the regiment. After three hours' rest,
they marched on Umballa, and arrived there at 7 a.m., May
15th, having accomplished sixty miles in thirty-eight hours!
A few days after cholera prevailed in the regiment epidemically.

The 2nd Fusiliers marched from the hill station of Subathoo
on the 14th May, 1847. At the foot of the hills they were
met by bullock carts, in which the men rode alternately.
They arrived at Umballa on the 17th, having left their
"bheesties" behind, and having been obliged to drink water

from stagnant pools and irrigated rice fields on the road-side. Cholera also prevailed in an epidemic form in this regiment.

The 75th *had only been one month* at Kussowlie, and marched forty-seven miles in two marches, and were also afterwards attacked by cholera.

It is stated by Marshall—and the statement is endorsed by every Indian medical officer—that excessive exertion in warm climates never fails to add to the sick list; and it is also well known that fatigue is a most powerful predisposing cause of cholera. Here we have soldiers, some of whom had been only one month in the hills, and must necessarily have been cachectic from former residence in the plains, suddenly called upon to make most excessive exertions in the hottest month of the year, obliged to pass through a district where epidemic cholera prevailed, marching in shirt-sleeves, without tents, carriage, water, or commissariat. As an almost inevitable consequence, these men suffered severely from disease; *and others would do so again, if called upon to make the same exertions under similar circumstances.* Such circumstances *may never occur again.* Although, in the eventful year of 1857 the State required the services of its troops at any sacrifice of health, money, or convenience, still the history of British India scarcely presents a similar emergency. On any future occasion, it is but rational to suppose that the exigencies of the service will at least allow of time being given for some preparation for the line of march, otherwise one of the chief desideratives, viz., the stability of the empire, which was expected to arise from the ruins of the cumbersome system of double government *will not answer expectations.*

Be this, however, as it may; let there be in future days parallel instances to the 1st and 2nd Fusiliers and the 95th Regiment; let numbers of men die from sudden removal from the hills for service on the plains; let cholera desolate regiments so moved—still I maintain that the mortality in any lengthened period, from such exceptional cases, would be

infinitely less than the common rate of mortality in regiments
on the plains!

*I am not, however, at all prepared to admit that every
regiment moving from the hills to the plains would suffer in
like manner.   On the contrary, I believe such would not be the
case, and it certainly would not be so in the cold weather.*

Moreover, there can be no doubt that Europeans descending from the hills, would take the field with an amount of
animal spirits *which they would not bring with them from
their stations in the plains;* their blood would be without
deterioration by heat and malaria, and although some of
course would fail—as would be the case in a European campaign—*the majority would require months to bring them to
that state of cachexia which they would have acquired in the
plains before the order for active service arrived.*

Dr. Morehead, in the remarks previously quoted from his
' Letter to the Secretary of State for India,' speaks of the
heavy sick list which would result from " seasoning fevers ;"
but a perusal of this talented physician's 'Clinical Researches '[1]
would lead any unbiassed reader to the conclusion *that there is
no such thing in India as seasoning fevers,* that "the general type
of disease in India, both in Europeans and natives, is asthenic,"
and that, as I have stated in this work and elsewhere[2] enun-
ciated, the European enjoys the best health during his first
years of residence in this country, and that his physical and
vital powers deteriorate in an inverse ratio with his length of
residence.   Granted, that " if it were possible to transport, in
a few hours, troops from the camp at Aldershot, to the plain
of the Ganges, use them for active service, and return them
in a few hours," a heavy sick list, and great mortality would
be the consequence; granted that the same would occur on
troops moving for active service, during the hot weather, from
the hills—*still I believe that the disease and death rate would*

[1] Morehead, ' Clinical Researches on Diseases in India.'
[2] The Author's ' Manual of the Diseases of India,' chap. i—iii.

be infinitely less than is the case at the present time, when our troops are moved out already deteriorated by heat and malaria, by residence in unhealthy stations on the plains.

If, however, the campaign occurred during the cold season, I am not prepared to admit that *any great sickness must necessarily result.*

But it is argued that, on the *conclusion* of the service, and on the regiment being moved *back* to the hills, " exhausted by heat, and fatigue, and sickness," the result would be " much mortality and invaliding from congestive inflammatory and organic visceral disease." It is further stated, " that the proof of this is no fancied picture, will be readily found in what takes place under the ordinary circumstances of troops fresh from England, arriving at the commencement of the hot season, and in what has taken place between the years 1840 and 1850, on the transference to the Himalayan hill stations of several European regiments, weakened by service, climate, and disease."[1]

It is, however, just that weakening, debilitating, and organic disease, which unfits regiments for residence on hill stations, which renders them prone to congestive and visceral disease. It is certain that this debility would not be induced by *a cold weather* campaign ; neither would active duty prolonged into the hot season debilitate a European regiment fresh from the hills to the same extent which would occur *if the men took the field with the degeneration half accomplished,* by simple residence in unhealthy stations on the plains. Hence, on return to the hills, Europeans who had formerly resided there, and only left for a time of duty on the plains, would not be so prone to diseases engendered by a lower atmospheric range as others who may be sent to the hills *already weakened by a longer exposure* or lengthened residence in an adverse climate, and therefore in the state

[1] Dr. Morehead's ' Report on the Sanitaria of the Poona Division of the Army.'

most liable to organic disease. It may be taken as a fact,
that the longer a European lives on the plains, the less
likely is he to receive benefit from the hill climates, for during
his residence on the plains he is daily in danger of contracting those diseases or tendency to diseases to which
sudden removal into a colder atmosphere becomes an obvious
and powerful aggravating cause. It also appears most probable
that men descending from the semi-tropical climate of
mountain ranges, where they would learn many desideratives
of Indian service, would take the field with far greater proba-
bilities of health than others could be expected to do, fresh
from Europe, and inexperienced in the every-day necessities
of tropical duty. In short, they would acquire experience
and habits of self-care, and therefore in this respect be better
fitted to withstand the effects of climate than if just landed,
ignorant, and heedless of exposure to the latter.

It is with the greatest deference I advance these opinions; but
holding them after mature consideration of the subject, under
all its various aspects, after attentive perusal of nearly all that
has been written regarding the effects of hill climates, and also
after some personal experience, I do not hesitate in avowing
my belief.

At the present period it is impossible to locate all our
infantry on hill ranges—political necessities will not allow
of this, however desirable the removal may be. Artillery and
cavalry, again, can never be quartered where difficulties of
carriage and access forbid. But it is not impossible to render
hill stations more available than they are at the present time,
and to send not only sick, but healthy men, to the less heated,
less malarious, and, therefore, more salubrious climates; instead of sending *only* the sick to hill stations, transporting
invalids, many in a state of suffering and exhaustion, I would
advocate sending *all* men. Thus, if a detachment of from
fifteen to thirty per cent., from every European corps—cavalry,
artillery, and infantry—were marched to the nearest hill

station, so as to arrive in February, and all those not requiring a winter in the hills, ordered to rejoin head-quarters
in November and December, the greatest benefit would result
to the greatest number. A seven months' residence in the
hills is sufficient to renew the *physique* of a man not organically diseased, and such men should be selected by the medical
officers of each regiment, with reference to their state of
health during the past year.

The only objection besides expense—*and the latter would
be quickly repaid by less disease and mortality*—which can be
brought forward against allowing all the European corps to
benefit in an equal degree by the hill sanitaria is, that the
men would suffer in their drill, or fall off in their discipline.
With *selected* commanding officers, however, this need not be
the case; and even admitting that there were temporary
deteriorations, and that the men returned to their corps a
little slack in their duties, better this than having to replace
them by raw recruits; better that they should appear a little
round-shouldered, with the ruddiness of health, than be
stretched out on hospital cots, and carried about in a dying
state in "doolies"—"better far to be in the hands of the
drill-sergeant than the doctor."

Such a plan as this might be carried out at the present
period; and in the course of time, as civilisation, and a better
knowledge of our home resources, increases among the natives
—as railroads extend over the country, and as more hill
ranges are explored,—a still greater number of Europeans
might be permanently stationed on the latter.

The remarks previously made naturally lead to a consideration of the peculiarities of hill climates. This, however, is
an immense subject, and cannot be satisfactorily treated of
in the present limits. Most hill ranges, most "islands in the
plain," have excellences peculiar to themselves, and it is only
a general approximation of these qualities which can be now
shown.

Dr. Frankland[1] observes:—"So intimate are the relations of man to atmosphere, that even the variations in the pressure of the aeriform mass surrounding our globe exercise a marked influence on the phenomena of animal life."

M. Von Muhry divides the atmosphere into three strata: The lower or hot region extending upwards 3,000 feet; the middle or temperate, extending from 3,000 to 12,000 feet, and the arctic above that heighth. In the first, terrestrio-miasmatic affections are prevalent in an intense degree; in the second, fluctuations of physiological phenomena are the characteristics according to the seasons—in winter, inflammatory, and tending to thoracic complications; in summer, congestive, and affecting the abdominal organs; in the third, none of the diseases first named, and a less amount of the latter,—frost-bite, scurvy, and loss of vital power from cold, being the phenomena observed.

In Indian hill-ranges, it is not the terrestrio-miasmatic causes of disease alone which are partially escaped; it is the absence of intense heat, the lowering of the temperature some 10 or 15 degrees, which allows the European to recover his elasticity of vital and physical powers—which the fervid heat of the plains depresses to the lowest existing point—which allows him to obtain a moderate quantity of exercise, without undue fatigue and exhaustion, and which conduces to his obtaining rest and sleep by night, free from the forbidding causes of the plains, heat and musquitoes.

As a natural consequence, the body not only becomes invigorated and inspirited, but the mind also is more active, and capable of greater and sustained action.

A clear intellect and a temperature of 88° Fahr. are almost incompatibilities, when long and continued mental exertions are required; and it may be stated that the capabilities of an individual as regards the latter, vary inversely as the heat of the climate. Intense application and deep thought have

[1] " Introductory Lecture, St. Bartholomew's ;" 'Lancet,' Oct., 1861.

never prospered for long together, where the body is always on the *qui vive* to keep itself cool. The hands, perhaps, may be taxed, but not the head. The latter, after a certain time, either fails, or performs its work unsatisfactorily.

Hence in hill stations, whether for good or for evil, there is generally a fresher, more energetic, and, it perhaps may be added, more intellectual tone than is met with amongst dwellers on the plains. There is not the heat to feel and talk about, and the climate seems to instil a new life into both mind and body. It gives a greater elasticity, and enables Europeans to undergo more than they could possibly endure under the "punkah," and "tatties," or exposed to the heat without those necessaries. It is the circumstance of a hill climate being a sanitarium *for the mind* as well as the body, which adds so immensely to its value.

Roughly speaking, the climate of hill ranges may be said to differ from that of the plains, in having a mean temperature of some 10 to 15 degrees cooler than that of the latter; in being above the range of the hot winds, and in the greater damp which exists during the monsoon season. Of course, various localities differ in minor points, the fall of rain for instance, on some mountain ranges being excessive, as at Mahableswar, where, in 1834, 297 inches; and at Nynee Tal, where, in 1853, 100 inches were measured. Again, in the Himalayas, a greater elevation will procure a colder climate; but the broad outlines of distinction are as just stated.

In all mountain stations the sun's rays are very powerful during the hot weather, and individuals are necessitated to recollect, that it is still the Indian sun which is above them. Hence the turban is required as it would be in the plains. Notwithstanding the power of the sun, the temperature is, however, considerably less, and hence those who in ascending neglect the precaution of warmer clothing, frequently suffer from an attack of liver, or diarrhœa. Natives from the plains who, either from negligence or want, do not adopt additional

garments, generally grow sick, particularly when badly shel-
tered, or indifferently fed, as is too often the case.

It has also been noticed by Dr. Mackay, on the Neilgherries,
by Grant, on the Himalayas, and by myself on Mount Aboo,
that salivation is easily induced. Hence in mountain climates,
mercury, if ever necessary, should be exhibited with caution.

Whatever may be the value of hill climates, it cannot be
too frequently reiterated that such benefit can only be very
partially obtained, unless sanitary matters are strictly attended
to at all hill stations.

A common-sense consideration of a hill climate, and a
comparison with the climate of the plains, will demonstrate
the fact that the former requires even *more* sanitary attention than any other locality either in or out of the tropics.

In temperate climates the putrefactive process is frequently arrested or mitigated by cold.

On the plains of India the same effect is produced by intense heat, the bodies of animals exposed to the glare of the
summer tropical sun scorching and drying up like mummies, *no decomposition taking place.*

As Jeffreys[1] observes, " Under the action of the weather,
where neither intense cold or heat retards the putrefactive
decomposition, this process goes on *viâ siccâ*, and, excepting
in the confinement of cesspools and sewers, of a character
by no means so injurious in its products as that which takes
place *viâ humidâ.*

Hence, in hill climates, where neither intense cold nor
ardent heat retard the putrefactive process, more remains to
be effected by sanitation than in any other locality.

Moreover, as would naturally be expected by any one conversant with the laws of disease, hill ranges are just those
localities where the worst diseases of both temperate and
tropical climes can exist—where neither heat nor cold being
ever sufficiently intense to destroy the germs of disease, the

[1] Jeffreys, 'The British Soldier in India.'

most favorable circumstances are presented for its generation.

Thus, Dr. Welb found that at Simla hepatitis both in the acute and chronic form is equally met with amongst Europeans and natives, and I have the same knowledge of hepatitis being prevalent on Mount Aboo. The cold of the mountain climate appears to disturb the portal system of the inhabitants of the plains of India, to render it torpid, and predisposed to congestive complaints. Dr. Welb also mentions that scrofula is very common at Simla, showing itself in tubercular deposit in the glands of the neck and axilla, in phthisis, and in caries of the spine.

In the year 1858, cholera in its worst form appeared at Munec at an elevation of 7330 feet above the sea level. The same disease was very prevalent at Dhurmsala in 1856 and 1857, at an elevation of 7000 feet. Dr. Baikie states, cholera has once occurred as an epidemic in the Neilgherries, at Coonoor, which is about 6100 feet above the sea level. Again, Dr. Balfour[1] states that in 1854 a very destructive epidemic prevailed at Coonoor, resembling cholera, but more fatal than that pestilence is usually observed to be.

Also Dr. Welb gives an account of typhus fever (generally found to prevail in temperate climates), which occurred in Simla in 1841. Nine years later, Mr. Grant witnessed typhoid fever in the same locality. Typhoid remittent was not unfrequent, some years ago, amongst the Europeans at Darjeling, and I also have seen isolated cases of continued and typhoid fever on Mount Aboo.

Epidemics of croup and diphtheria have latterly occurred on the hill ranges.

Smallpox has raged with great virulence amongst the inhabitants of most hill regions in India.

I believe it may be stated that malarious fevers, intermittent or remittent, originate *occasionally* on all hill

[1] 'Localities in India exempt from Cholera.'

stations in India, in individuals who never had fever before; and relapses continue to occur for years. These fevers are, however, of a milder type than others of the same nature, originating on the plains.

Colds, coughs, and pectoral complaints are more prevalent on the hills than on the plains below.

But, in addition to the diseases of both temperate and tropical climates which have their chief locality of union on hill ranges, the latter have also diseases *sui generis*, and which only originate in such places.

Reference has already been made to the " Mahammurree," or Indian plague; by some described as a malignant fever of a typhus character, accompanied by external glandular tumors, and by others regarded as the true plague—which, having commenced its ravages near Kedarnath, in the snowy range, traversed the Himalayas, from thence downwards.

There is, however, another disease which occurs with more or less intensity in most hill climates; I allude to "hill diarrhœa," or "diarrhœa alba," a disease characterised by painless diarrhœa, occurring generally in the mornings and evenings—copious, white or slightly yellow evacuations, and progressive wasting or cachexia, from which the patient ultimately sinks. This disease is so prevalent at Simla as to have acquired the vulgar name of " Simla trots." It is also frequently seen at other hill stations, and a full account of it may be found in my work, entitled, ' A Manual of the Diseases of India.'

A consideration of the various maladies I have mentioned as prevailing in hill climates, will demonstrate the fact that the great majority of such diseases are of the zymotic variety, and therefore preventible, or at least mitigable.

Conservancy, for the reasons previously advanced, requires more attention on the hills than in the plains below. As is shown in the chapter on this subject, the means of conservancy most applicable to India is manual labour; and to this end men should be employed with bullocks and panniers, where carts cannot be used, to carry away all ordure and other refuse matter. This is a system I have vainly endeavoured to establish on Mount Aboo, but, until the present day, the refuse of each house is thrown somewhere in the locality, and this will ever be the case until public scavengers are entertained. As I stated in my 'Annual Report for 1861,' until this is effected, the inhabitants of Mount Aboo will not be on a level, as regards this essential point of civilisation, with the Israelites of old, who, under the inspired Mosaic laws, were obliged to remove and bury their excrete.

In all hill stations, public latrines should be erected for the use of the natives, and should be cleansed at stated periods by public scavengers. Otherwise—as is the case at Aboo, as is the case at Ootocamund, as was the case at Simla—the natives make use of every bush and ravine, as temples of " Cloaca," and heap dirt *stratum super stratum*, the stench of which rivals the smells described by Coleridge as existing at Cologne.

The entire station and its neighbourhood should be cleared of deciduous vegetation, and while trees are religiously preserved, the growth of the former should be prevented.

Of still more importance is the subject of drainage; and regarding the necessity of this work much ignorance exists, many imagining that such means to carry off water is not required on a mountain elevation. That drainage, however, on hill ranges can hardly receive too much attention may be quickly proved.

Mountain ranges consist of rocks, in the valleys and hollows of which, during the lapse of ages, a rich alluvial soil has collected, formed from the crumbling and disintegration of the rocky sides, and the decay of vegetable material. This earth is exceedingly porous, and retains water like a sponge, which, moreover, cannot drain from or percolate through the rocky depressions and cavities into which it gravitates; hence, under a tropical sun, this moisture is constantly drying, such process being attended, as is the case in every other locality, by the extrication of malaria.

From the reasons already advanced it must, however, be evident that sanitation is even more requisite in the hills than in the plains. Unless clearing, draining, conservancy, structure, position, aspect, and locality of buildings, are made matters of chief importance, our stations in the hills will be merely *refuges from heat during the hot months, and sanitaria only in name.*

The site of a hill station should be fixed, if possible, on the windward side of the mountain range, the summit of which *should not be crossed;* otherwise, as is the case with Aboo, the the station will be situated in a basin or hollow, thorough perflation of ravines will not be obtained, and drainage will be rendered most difficult, if not impossible.

If a station can be situated, as Poorundhur, for instance, on the declivity of the range, it receives the air fresh from a boundless expanse of atmosphere. If, on the other hand, the locality is fixed on the leeward side of the range, or in the centre thereof, the air, in passing through the intervening ravines and jungles, becomes impure and malarious, and very different from what it was when it first struck the mountain-top.

Sir Ranald Martin[1] observes :—"As to barracks on hill stations, provided the site be of sufficient elevation, well cleared, drained and levelled, with a good water supply, the material need not be of a costly character, well-constructed huts for the accommodation of ten or twelve men forming a good protection against the inclemency of the weather. Such simple and cheap structures will, I believe, prove more conducive to the health of the European soldier on the mountain-ranges of India, than the most costly barracks ; and the same may

[1] ' Letter to Chairman of Court of Directors,' 1857.

be said of hospital huts for the reception of the same number of men."

Sir Ranald then quotes the experience of European campaigns, where soldiers who were hutted fared well, and those massed in barracks or hospitals perished at enormous rates; also the fact of the Irish typhus fever patients perishing in large numbers in the workhouses and hospitals, and recovering along the road and under the hedges in goodly proportions; also that Sir Charles Metcalf built only huts in the mountains of Jamaica, and that Indian experience shows that costliness of materials will not prevent disease.

With the greatest deference to this high authority, and to others who hold the same opinion, I venture to express my conviction, that residences intended for Europeans in Indian hill climates should be erected, in most essentials, as though intended for other climates. In the cold season of the hill climate, although the rays of the sun may still be powerful, the interior of dwellings are considerably colder, and it is a common practice of soldiers and others to leave the interior of their barracks and bungalows, and sit out in the sun and warm themselves. Again, when the sun sinks below the horizon the cold becomes intense; hoar frost generally covers the ground, and in the Himalayas snow may fall. Hence, if the soldier is not in a position to defend himself from the inclemencies of the weather, the diseases of colder climates—such as pectoral affections and scrofula—are apt to develope themselves. Also, as has already been stated, the heat during the hot weather, without being intense, as it is in the plains, is frequently excessive; and for this cause a thick but not too massive building—as the latter would absorb heat and radiate again at night—is essentially required. The arguments and similes brought forward by Sir Ranald Martin are, moreover, scarcely applicable to hill stations. It is beyond doubt that, were crowding of human beings to take place in India hill climates, as occurred in barracks and general hospitals in the former European campaigns, or as happened during the Irish famine and typhus epidemic, disease and great mortality would be the result; *but crowding ought never to occur in Indian sanitary barracks and hospitals;* and if this be the case, such arguments carry no weight. On the other hand, the diseases which arise from exposure to cold, or from residence in low, draughty and leaky huts, are self-evident; and although these diseases are not so virulent as others arising from overcrowding, still it is advisable to erect dwellings equally free, and to keep them equally free, from both the latter and from the former sanitary defects.

As a rule, with few modifications, the barracks and dwellings best calculated as residences for Europeans on the Indian plains will be found to afford the best means for the preservation of the health of the same classes on Indian mountain ranges.

The position and aspect of barracks on hill stations is not of that paramount importance which it should be considered on the plains. From the peculiar formation of mountains, it is frequently impossible to obtain a space of level ground sufficiently large to allow of the erection of barracks *fronting* the prevailing breeze—*a desiderative which ought never to be neglected in the plains.* In elevated regions, however, on account of the diminished temperature, and the greater coolness of the breeze, barracks, &c., may be erected at considerable angles to the quarter from whence the wind blows; and although I cannot conceive it possible for any person to recommend the erection of barracks gable-end on to the wind, as the barracks at Aboo at the present time are situated, still the ground may often only admit of a diagonal direction being taken. This, however, in hill climates will permit that ventilation and perflation which it would prevent in the plains.

Regarding the position of buildings, little need be said.

With strict sanitary regulations there would be no danger of human residences being erected near or over marshes, as is the case on the Neilgherries, and at other hill stations. Strict sanitation would prevent this, by the simple drainage of such malaria-generating spots, even should any one be sufficiently perverse to erect his private residence in a naturally unhealthy locality. For barracks and public buildings the driest and most elevated spots should be selected, and valleys avoided. The idea formerly prevalent, and which appears to have been acted upon at Mount Aboo, that an overhanging rock or mountain shelters the barracks from too cold breezes, should be utterly abandoned. The evil of placing barracks in a valley, overhung by a mountain 1000 feet high, is shown by the amount of fever which originates in the Aboo barracks. The advantage of placing the men on one of the most elevated spots on the range has been equally forcibly demonstrated during the last season, when, by my advice, the soldiers left the barracks and camped on the last-named elevated site.

The proper height at which to establish a hill station has been a subject of much discussion, and even now may be considered as *sub judice*. Many have held the idea that malaria ceases to exist at a certain height above the level of the plain, but that at a little greater elevation the European is brought within the range of another class of diseases, namely, that of bowel complaint. The first of these theories is still more erroneous than the latter, for the liability to fever depends, as cannot be questioned, more on the character of the soil and geological substratum than on height or any other circumstance (for remarks on this point see the chapter on "Malaria.") That the latter idea is not correct may be inferred from the facts that at Mahableswar, Mount Aboo, and on intertropical hill ranges generally, bowel complaints do not occupy a prominent position, and may frequently be traced to neglect of suitable clothing, or, as is probably the case at Simla, to the dirty condition of the ground.

Height, then, that is to say, any specified height, such as six, seven, or eight thousand feet, above the level of the sea, is not a *sine quâ non* for a hill station. It is sufficient if the elevation removes the European from the stratum of hot air, and, therefore, above the influence of the fiery winds. If the height be sufficient to effect this, it is, I believe, nearly all that is required in this respect. Saussure's theory was, that temperature decreases one degree in every 350 feet of ascent, and an *abrupt* elevation of about 5000 feet will not only reduce the temperature sufficiently, but, moreover, be quite high enough to escape the heated winds of the plains. If the mountain range does not present an abrupt elevation, and undulates gradually into the plains, it will, for obvious reasons, have a higher thermometric scale than the reverse, and probably require a longer ascent.

Such chains are, however, not well fitted for sanitary stations. These undulations at the base of mountains are always unhealthy, and where either these or a " terac" exists, the ascent and descent of the mountain will always, at certain periods of the year, be fraught with danger. I do not, however, understand why, as some have stated, a mountain range bounded by a terac should be unhealthy *in itself*, if the elevation were sufficiently abrupt, and high enough to prevent malaria being rolled from the terac on to the summit.

The question of locating the supreme government in one of the Himalayan stations has been latterly mooted, and a few words may, with propriety, be here devoted to the subject. It is more important that the high officers of state should retain their health than any other members of the community. When the reverse is the case, their work stops, as it cannot, like an officer's, be done just as well by any one else, simply because there is no one to do it. Meanwhile, though their work stops their salaries go on ; so that, on the mere vulgar grounds of expense, their inability to superintend the machinery of which they are directors is a most serious loss. Again, a clear intellect is more especially necessary for the rulers and heads of departments than for any other classes; and it has already been demonstrated that this is incompatible with continued exertion and the heat of the plains for any lengthened period.

There are, however, objections to making any of the Himalayan stations the seat of government. Such a plan would place the ruling power a thousand miles away from the sea, which has always been the origin, and must continue to be the seat of British supremacy in India. We have conquered India from the sea; and if we leave the ocean, we endanger the continuance of our hold on the country; "for if our communications with England are maintained, we could always reconquer India, even if it were for a time wrested from our grasp."

The capital of India, then, ought to be a sea-port, in the immediate vicinity of which a mountain retreat might be found. Calcutta is only approached by a journey of many miles up the Hooghley river, and has no hill sanitaria nearer than the Himalayas; Madras has no harbour, and is therefore unfitted for a large commercial emporium; Bombay, however, has both a most commodious anchorage and the hill stations of Matheran and Mahableswar, at a distance, respectively, of a couple of hours and a day's journey, and both connected with the city by railroad.

In addition to the great political advantages of Bombay, as possessing the finest harbour, as being the nearest point to England, and the natural thoroughfare to all parts of India, there are sanitary excellences peculiar to the locality. These consist in the sea breeze, which is always ventilating Bombay, and prevents that muggy, stagnant atmosphere so prevalent in Calcutta, in the proximity of the comparatively healthy table-land of the Deccan, where, according to Dr.

Morehead, the rain season of the south-west monsoon is less inimical to the European constitution than in most other parts of India, and in the exemption of the hill stations named from a fever-belt or terae at the base, of the danger of passing through which at certain seasons the lamented fate of Lady Canning furnishes sad demonstration, her death being understood to have been occasioned solely by malarious fever, contracted in the jungle which occurs in the descent from Darjeeling.

CHAPTER X.

ON MARINE SANITARIA.

Diseases not benefited by Hill Climates—Diseases benefited by Sea Climates—
Peculiarities of Sea Climates; Advantages: Moist and Saline Atmosphere;
Iodine; A Fish Diet—Coasts to be avoided—Locality to be chosen.

It has been demonstrated in the foregoing chapter, that
the curative powers of hill climates are extremely limited, and
also that there are many diseases over which the atmosphere
of the elevated regions exerts even a baneful influence. Such
affections are diseases of the lungs, particularly scrofulous
changes or consumption; dysentery, diarrhœa, and indeed
all disturbances of the bowels; liver diseases; head affections;
neuralgic complaints, if not malarious; malarious fevers,
during the cold season; and generally all cases of rheumatism.
The removal of a person affected with lung disease, such as
tubercle, or vomicæ, into the ten or fifteen degrees less heated
atmosphere of the mountains, is frequently followed by acute
congestion, and fatal hæmoptysis. In dysentery, liver disease,
head affections, the same congestion of internal organs,
caused by the sudden application of cold, cannot fail to aggravate any disorder which may be present in a diseased, *and
therefore weakened* part; while rheumatism, the first or cold
stage of ague, and neuralgic remedies, as every one knows,
are but very rarely (the second affection never) benefited by
cold.   Briefly, it may be accepted as a truth that all cases of
organic, or threatened organic disease will not be benefited
by the hill-climates of India, and it is in such instances that
the value of removal to the sea-coast becomes apparent.

The influence of the sea on climate is of vast importance, and arises from several causes, the principal of which are as follows :

The first difference, viz., the greater uniformity of climate on the sea-coasts, depends substantially on the peculiar physical properties of the water and land. The former having a great capacity for heat and a feeble conducting power grows warm only slowly from the power of the sun, while the latter rapidly absorbs the solar rays. From the same physical properties, the superficial layer of water, being cooled by evaporation, becomes heavy, sinks down, and gives way to the warmer molecules of the inferior strata, while the soil, since it cannot be so displaced, is not only heated, but also cooled in a considerably shorter space of time.

It will, therefore, be evident that, although the sun may shine on both sea and land, the sea must be colder than the land during the day, and warmer during the night. In the same manner, taking the different seasons of the year, in summer the sea is colder than the land, in winter it is warmer ; it preserves the mean temperature, while the land experiences the extremes ; and the air of the sea-margins, sharing in the uniformity of temperature which belongs to the surface of the waters, helps to give the sea climate its peculiar characters.

Air in motion, however, is always cooler than stagnant atmosphere, and hence the land and sea breeze combine in lowering the temperature of the coast several degrees below that of inland districts. In proportion as the sun rises above the atmosphere, the land becomes warmer than the neighbouring sea. Their respective atmospheres participate in these unequal temperatures. The fresh air of the sea rushes from all directions, under the form of a sea-breeze which makes itself felt along the whole coast, the warmer and lighter air of the land ascending into the atmosphere. During the night the reverse happens ; the land loses heat and cools more rapidly than the sea. Its atmosphere, having become heavier, rolls into that of the sea, under the form of a land breeze, and this lasts until the temperature and density of the two atmospheres has again become the same.

Thus with *uniformity* of climate we have combined air in constant motion, and a temperature slightly less than that of the interior of continents, all of which are well calculated to benefit the cases of disease previously mentioned.

These, however, are not the only advantages which a sea climate affords. The moisture of the atmosphere and its impregnation with saline particles and iodine, is frequently most beneficial to the invalid who has resided in the interior, subject to the influences of the hot, arid, and parching winds; a removal into the moist atmosphere of the sea-coast is a change which invigorates both mind and body; while in particular cases of disease the saline atmosphere and the iodine given up from the water to the air may be expected to exert a beneficial effect. In short, "the sight and sound of the ocean are as refreshing to the exhausted spirit as the breeze that blows from it is to the exhausted mind. The air from it is redolent of health."

In addition to the effects of sea climates over disease there is another advantage which the invalid may desire from the locality of the ocean, and that is, a daily supply of fresh fish. To the sick man from the interior, whose appetite has failed and whose convalescent diet has consisted of indifferent compounds prepared from poor meat, and farinaceous food, the value of a supply of fresh sea fish, is great indeed, and may in many instances prove the means of a speedy recovery of health, instead of a long and tedious convalescence. Sea bathing may also be beneficial in very many instances.

Although these advantages are to be derived from a sea-coast residence, still it is not every ocean boundary which will afford them. Some coasts are low-lying, flat, have immense surfaces of mud and decaying seaweed uncovered at the ebb-tide, permit no vegetation for miles inland, and are otherwise so situated as to be not only unadvisable residences, but also unhealthy. It is in such low-lying coasts that scurvy (see Chap. XX), beri-beri, leprosy, and kindred diseases are endemic, and hence much care will be required when fixing on the site for a marine sanitarium. All arms of the sea and gulfs should be avoided, the blue and open ocean in front being a *sine quâ non*. Low, flat, sandy coasts are most frequently unhealthy in the tropics, and, therefore, not to be recommended, and, of course, the same remark applies to all localities near the mouths of creeks and rivers.

A rocky shore, with *elevated* cliffs, and the open ocean in front is the *only* locality which can be recommended for a marine sanitarium, and if such a position, from jutting out into the sea is surrounded on both sides by water, it will be an additional desiderative, reducing, as it must do, providing no large surface is left exposed, the formation of malaria to a minimum (see chapter on ' Malaria'), and rendering the temperature and moisture of the atmosphere still more agreeable and uniform.

Of course the surface of the ground composing the shore and the surrounding country, its capabilities for drainage, and freedom from marsh and jungle, and from overhanging mountains, with the water supply, and physical condition of its inhabitants must be taken into consideration when fixing the site of a marine sanitarium, according to the same rules laid down as applicable to stations on the plains. (See Chap. XI.)

Dr. Macpherson has latterly recommended the island of Martaban, which, from its geological characteristics, size, and proximity to both Madras and Calcutta, would appear admirably adapted for a marine sanitarium. Islands, however, unless well raised from the surface of the sea, are not generally healthy, and it is essentially necessary that such localities should be observed during the winds, damp, and fog of the monsoon, as well as in the fairer seasons of the year.

CHAPTER XI.

ON THE CHOICE OF STATIONS IN THE PLAINS.

Stations in the Plains can never be wholly abandoned—The Spleen-test of Locality—Proximity of large Rivers—Change of Course of Rivers—Mouths of Rivers to be avoided—Dry Beds of Rivers to be avoided—Sand not Dry—Soils to be avoided and chosen—Water—Cultivation—Drainage—Canal Irrigation—Uneven Countries condemned—Proximity of Mountains condemned—Neighbourhood of Cities condemned—Burial Grounds—Prevailing Breeze—Model Cantonment—Time to choose the Site—Conclusion.

As time develops the capabilities of hill climates in the prevention of disease among Europeans, it is more than probable that stations and cantonments in the plains will be gradually deserted for mountain sites ; but even in this case, political necessity will always render it impossible for our troops to vacate the plains altogether. Moreover, the difficulties of carriage, and the height and inaccessibility of mountains are such as to forbid the hope that artillery and cavalry, especially the former, can ever be, excepting in exceptional instances, located at hill stations. Hence, a short consideration of the desideratives of cantonments on the plains must be undertaken.

The most important point to take into consideration in fixing the site of a new station, is the condition of the inhabitants as they appear when engaged in their every-day occupations. Are they a robust and healthy race ? As Dr. Dempster truly observes, " To an eye accustomed to the various races and castes of India, the aspect of the inhabitants of any particular district will convey useful information, provided always the inquirer does not deceive himself by any preconceived opinion."

In 1845 a Committee was appointed to report on the cause of the unhealthiness of Kurnaul, and of the portions of the country along the line of the Delhi canal. It was suggested by the medical member, Dr. Dempster, to examine the spleen, a course which should never be neglected, as almost *the most* important method of determining the healthiness or otherwise of any given locality in a malarious country.

In such districts where intermittent or remittent fevers are prevalent, a certain proportion of the population will be found affected with chronic enlargement of the spleen—a disease which, either primarily or secondarily, is a certain manifestation of the presence of malaria, and has been well designated by Dr. Dempster as the *Experimentum crucis* of the whole investigation.

Where this disease is found to exist in a large proportion of the inhabitants, the locality ought to be unhesitatingly condemned as a cantonment for European troops ; for although the district might possibly be *made* healthy by drainage, clearing, and other sanitary regulations, still it may be stated as a fact, that such procedure would entail an enormous outlay of the public funds.

The presence or absence of spleen-disease having been ascertained, as the most important matter, the proximity or otherwise of large rivers should next be inquired into. It is a well-known fact that Indian rivers, flowing through alluvial soils, change their course, generally to leeward, in the course of years ; and hence it happens that many river stations are bounded in the rear by a creek, or chain of shallow pools, representing the old course of the river. The destruction of the ancient city of Gour has been referred by Dr. Chevas[1] to this cause ; and the towns of Dacca, Mymensing, Hooghly, Nuddea, Dinapore, Berhampore, and some others, are cited

---

[1] Chevas, ' Ind. An. Med. Sci.,' vol. xi.

by the same author as occupying unfavorable situations of this kind.

The mouths of rivers, and indeed within the tidal influence, are especially to be avoided, as the localities for cantonments. Most Indian rivers demonstrate this by the number of mouths through which their contents are poured into the ocean. The Ganges, the Indus, indeed all the large rivers, present numerous ancient channels to the sea. Lyell,[1] in his account of the delta of the Ganges, states there are eight openings, each of which has evidently, at some remote period, served in its turn as the principal channel of discharge.

It is not, however, these large rivers alone which should be avoided, and it is not merely the facts of their having in former times changed their course, and left a tract of marsh or nullahs, which renders their locality unhealthy. All rivers which in the dry season become comparatively narrow watercourses, and which, therefore, leave on either side an enormous tract of sand, are essentially malarious, and therefore, like the damp and drying banks of larger streams, unfit the locality for the residences of European troops.

An opinion is very prevalent that sand is dry, and it not unfrequently happens that tents are pitched on so-called dry sandy beds. Removal of a few shovelsful of the dry-looking surface would soon demonstrate the reverse; and, indeed, the very individuals who place their tents on such places may frequently observe their servants digging to obtain the water underneath. Moreover, all decomposable matter which the floods deposit upon the sand, "not finding any neutralising agents there, undergo disintegration on the surface, and pollute the atmosphere." The evil results of placing stations too near to large rivers have been demonstrated over and over again in India, and many cantonments have been deserted from this cause. Thus Loodianha, now forsaken as a European station, was situated on the banks of a water-

[1] Lyell, 'Principles of Geology.'

course, formerly a channel of the Sutlej, which now flows seven miles to the north! Berhampore, where the annual mortality was at the rate of 90·69 per thousand, was deserted in 1826, the cantonments being placed between the Ganges and a jheel formerly the old bed of the river. Ghazeepore, Cawnpore, the forts of Allahabad and Agra, are all situated immediately on the verge of rivers, and all have from time to time proved most unhealthy for European troops.[1]

The next point to be ascertained is the nature of the soil, and geological characteristics of the district. Stiff clay lands, and such as are retentive of moisture, are not fitted for the location of troops. Light, friable, and gravelly soils, on the other hand, are well adapted for cantonments. Lands, however, which, although gravelly on the surface, present a substratum of clay at a very moderate depth, should not be chosen. This, however, although occasionally occurring, is not a very prevalent geological characteristic in India. Soils having *laterite* for a foundation are frequently healthy, and often escape epidemic attacks of cholera; while those consisting of alluvial deposit, resulting from the decomposition of *trappean* or *metamorphic* rock, are generally the reverse, and especially liable to the latter disease.

It has been noticed in the chapter on malaria that ferruginous soils, and indeed all grounds destitute of vegetation, are essentially malarious.

A most important consideration is that of water, a plentiful supply of which is, of course, a *sine quâ non* in the immediate locality of a cantonment. As Mr. Dempster remarks, "good water to a native of India comprehends everything we express by a perfectly healthy locality, and therefore the inhabitants of the district should invariably be questioned on this important subject." For further remarks on this matter, and for the chemical and ready methods of testing water, I refer to the chapter under that head.

[1] Dempster's ' Rules for the Selection of Sites for Cantonments.'

The amount of cultivation and the capabilities of drainage must next be taken into consideration. There are few localities which will not allow of the former being carried out, to a greater or less extent; but it is incumbent to fix the position of a cantonment in that spot where the natural slope and formation of the ground will leave the engineer but little to effect in the matter of drainage. By so doing, both the liability to disease will be most materially lessened, and also a source of original and continual expense will not be incurred. As a general rule, open drains of a rudimentary description should be adopted. A shallow surface-drain, says Dr. Hathaway, made on the ground with bricks, is the " best means of carrying off rain-water, and it can be effectually cleansed or renewed from time to time without expense." It must never be lost sight of that marshy spots formed by *simple water* induce intermittent fevers and their consequences, as well as other circumstances more particularly noticed under the head malaria; while want of drainage of foul water and filth localise and spread epidemic diseases, and cause in the inhabitants who live near them a proclivity to such affections, and a general adynamic state of constitution, or, in more familiar terms, a state of health " below par." Whatever locality, therefore, presents insurmountable obstacles to *thorough drainage*, whether from the configuration of the surface or the peculiarities of the soil, should be rejected as the site for the dwellings of human beings.

The presence or absence of cultivation will in many countries, particularly in tropical and semi-tropical regions, exert a great influence over the health of the inhabitants. Experience has again and again demonstrated that malarious disease will arise from a drying surface of moist earth; and therefore irrigated lands, such as rice-plantations, in the neighbourhood, and especially to windward of stations, cannot be too strongly deprecated. If it were possible to irrigate cultivated fields, supplying them with just sufficient water to promote the growth of the crops and no more, the malaria arising from such grounds would be comparatively small. Hence gardens which are not profusely irrigated do not prove deleterious in the neighbourhood of residences, and indeed may, under certain restrictions, be considered salubrious. If, however, these same gardens were as extensive, and flooded with water in the same manner as rice-crops are; if stagnant puddles were allowed to exist in gardens, as is the case in cultivated lands; and if gardens were situated in valleys, or at the tail of ponds, as cultivated lands are frequently placed, the former would prove as inimical to human health in tropical regions as the latter.

Canal irrigation will always prove more unhealthy than irrigation from wells, simply because the supply of water is more abundant, and more readily obtainable from streams which communicate directly with a noble river; and thus the ground becomes more saturated, and a large quantity of stagnant puddles result. Cultivation in India being frequently only limited by the difficulty of obtaining water, it follows that in the neighbourhood of canals the greatest surface of cultivation must exist, which, as above stated, is another cause of increased malaria. Lastly, artificial canals are sometimes little less than embankments conveying the water over the country *at a greater elevation than the surface of the soil.* Where this is the case, any kind of drainage is quite impossible, either natural or artificial, and malarious disease is therefore rife.

A country abounding with shallow ponds, jheels, and nullahs must be condemned as the locality for a cantonment, as such surfaces can scarcely be drained without the expenditure of enormous sums.

The immediate proximity of mountains must also be condemned, whether the elevations are of sufficient magnitude to transmit water enough during the rains, to keep the country at their base in a perpetual state of moisture, and so give rise to a ferae; or whether they are merely small rocky formations, capable only of intercepting the breeze or radiating the heat. The unhealthiness of a ferae does not require demonstration again, and any one who has lived in a cantonment surrounded or overhung with black rock, will willingly admit its absorbing and radiating powers.

The immediate neighbourhood of large cities, especially in the leeward direction, are not satisfactory localities for our troops. The sanitary condition of all Indian towns cannot be described as less than vile, and the situation of many has been chosen more on the score of convenience, as being near a river, &c., than from any other cause. Former political necessities led to many of our cantonments being fixed in the immediate locality of large cities, and their unhealthiness has already been referred to. In other respects also, as allowing of greater intemperance and immorality, their proximity should be avoided.

Old grave-yards must be shunned altogether. This would appear a superfluous caution; but it has happened, as at Sukkur, that Indian cantonments have been located either on such surface or in the immediate locality. In like manner, new cemeteries should neither be established in the centre or to windward of stations, but at some distance to leeward. The same remark applies to Hindoo places of incremation.

In this country—Western India at least—the breeze continues during the greater portion of the year blowing from one direction, viz., south-westerly; but we find that some of our military cantonments are fronted to the east, and have the refuse of the camp placed in rear, that is, to the westward, from which the breeze blows. Not only is this the case, but outhouses, such as cook-rooms, servants' rooms and stables, are often found also between the wind and the dwelling-house; and worse even, if possible, as a general rule, the bath-rooms and necessaries will be found occupying the same relative position within or at the sides of dwellings. A recent writer in the 'Times of India' (September, 1861) asks, under such arrangements, " How is it possible that pure air can be obtained ?"

The same author observes :—

" Did space permit of it, a table of the monthly winds would prove equally instructive and valuable. The great burial-grounds at Poona are so placed as to be between the camp and the two most prevalent winds, and in that direction more than half the refuse and excrements of the camp used to be deposited. Yet, with this and other malarrangement, Poona, owing to its naturally salubrious position, has been a healthy station for Europeans. How much more so it might have been, had the great value to health of the pure westerly breeze been understood and recognised, it is impossible to say. Doubtless, the elevated plains of the Deccan are naturally the most healthy locality within the tropics for European troops, unless it be on the tops of the hills, and I believe we may safely set down half the mortality that occurs there to the want of sanitary science, as evidenced in the bad arrangement of almost everything affecting the health of our cantonments."

All the foregoing unadvisable concomitants being avoided, an Indian cantonment would be placed at a distance of some miles from and to windward of both large rivers and cities. It would be located on the summit of a gentle un- dulation, or on the slope of a rising ground, not sufficient to overhang the station, to intercept the breeze, or to radiate heat. The surface would be light, friable, or gravelly soil, without a substratum of heavy, retentive material. Water would be plentiful, to be obtained at a moderate depth, and of good quality, according to the requirements and tests laid down in a future chapter on this important subject. A country intersected by nullahs, abounding in jheels and tanks, and liable to flooding, would be avoided. Localities not susceptible of easy and perfect drainage would be shunned, and old grave-yards and the sites of ancient cities also escaped. The whole cantonment would front the prevailing winds ; native houses, bazaars, burial-grounds, places of Hindoo incremation, in fact, all essentials, excepting wells, would be placed rearward to European residences.

In addition to these requirements, the station should be surrounded by a zone of one or two miles radius, free from cultivation and irrigation. Such extent of country should be grass land with clumps of trees or occasional gardens ; sanitary matters should be strictly attended to throughout the whole extent, and all refuse from the cantonments should be removed by manual labour without these limits, if not susceptible of being destroyed by fire within such space.

A cantonment situated as above described, in which sanitation was carried out, not only with respect to removal of filth and débris, but also as regards the structure, aspect, and requirements of buildings, would give the European as great a chance of escaping the diseases and deterioration of health inseparable from residence in a tropical climate as the class from which the rank and file are at present drawn enjoy in their best aspects in most portions of the United Kingdom, of escape from the ills which *their* lot is there heir to.

Dr. Dempster very properly insists upon the necessity of choosing new sites for stations, not in the dry and cold season when all appears clear and dry, but in, or immediately after the rains, when the defects of locality, as regards natural drainage, the presence in the vicinity of pools, marshes, or water-courses which overflow their banks and flood the adjoining levels, together with the presence or absence of rank, deciduous vegetation, which is eaten up by cattle, or otherwise disappears during the cold season, may be readily and unmistakably ascertained.

In conclusion, I would wish to see, in the laying out of any new cantonment, the gymnasium commenced as soon as the parade ground, the cricket field ordered as well as the " conjee " house, and the soldier's garden sanctioned as soon as the canteen.

CHAPTER XII.

ON CLEARING.

It is admitted by all sanitary authorities that in tropical climates the entire neighbourhood of barracks and hospitals should be cleared of all jungle at any cost of money and labour, so that the locality may be free from all sources of miasmata and all cover for accumulating filth and decomposing matter.

From the number of human beings who inhabit barracks and hospitals, clearance is of more importance near such buildings than around residences giving a larger cubic space to their occupants.

Clearing of deciduous vegetation is, however, sufficiently momentous to render it incumbent that the limits of all Indian stations should be completely freed and maintained clear of all such causes of disease.

By clearing is not to be understood the cutting down and destruction of large trees, which, on the contrary, should be carefully preserved. The grateful shade they afford, the pleasant green they present to the eye; the collection and condensation of vapours, and the equalisation of rain-fall they induce; the less rapid evaporation, and consequent greater coolness resulting from their presence; the prevention of the ground becoming so heated as their absence would allow; the probability that springs draw their supplies from sources in the immediate vicinity of trees, and that their absence dissipates the water, are cogent reasons for their preservation.

It is however, necessary to exercise discretion as to number, and more especially as regards position. Thus, they should not be allowed to intercept the prevailing breeze; they should not be placed in immediate proximity to, or allowed to overhang inhabited dwellings, and their spreading boughs should be trimmed to within seven or eight feet of the ground. If permitted in positions where they intercept the prevailing breeze, they deny thorough perflation and ventilation; and moreover, air passing through the thick foliage of many Indian trees, becomes more or less charged with organic impurities, and enters the dwelling in a less satisfactory condition than if it had met with nothing to obstruct its onward course. The odour from some trees is also very disagreeable.

The power which trees exert in retaining malaria amongst their foliage (see chapter on Malaria,) is another urgent reason against their being allowed to overshadow a tropical residence, as, besides danger of malaria passing *through* the tree, any person going in or out would be exposed to miasm, attracted by the foliage above, and falling down from its inherent specific weight before it could be consumed or decomposed.

The probability, also, that trees inspire oxygen by night, or at all events the fact that they cease to inspire carbonic acid, is another reason against their proximity to dwellings. By consuming oxygen—so rarely in excess in our rarified Indian atmosphere—they lessen the amount present in a stagnant air for the respiration of human beings, and by ceasing to inspire, or even by exhaling carbonic acid, they, equally with animals, help, during the absence of sunlight, to render the atmosphere impure.

Again, as Dr. Chevers observes, " Trees, like men, have their periods of accidental sickness and of natural decay, and an unhealthy tree at our chamber window is certainly no desirable companion during the nights of an Indian autumn."

Some trees would appear to attract malaria to a greater extent than others. In most parts of India the dense foliage of the tamarind and nimb tree is dreaded by the natives as a canopy during the night; indeed, it is a popular saying amongst the inhabitants of Bengal, " that he who has a tree near his house keeps death at his door."

Thus, although the absence of trees is a certain cause of disease (see chapter on Malaria) in malarious localities, still their presence in any improper position is equally to be deprecated.

With deciduous vegetation, however, the matter is different. " Mudar grass," gigantic convolvulaceæ, with their sickly odour, the prickly pear, or any species of cactus, should be ruthlessly rooted up and destroyed. Their presence are frequently signs of undrained ground ; their decay retains malaria, and pollutes the atmosphere, and they are the abode of numerous insects and reptiles, which live, die, and decay amongst their stems.

CHAPTER XIII.

ON BARRACKS.

DR. CHEVERS states, " The highest, clearest, and driest spot in a contonment—that best exposed to the prevailing currents of air, and most remote from any permanent source of malaria—that which a Roman general would have taken for himself and his prætorian cohort—should be selected, regardless of all other considerations, for the barracks and hospital."

The histories of very many European regiments during their service in India incontrovertibly prove that a large proportion of the excessive sickness and mortality which prevails among soldiers in India, is traceable to defects in the site, aspect, construction, or sanitary economy of the residences they have occupied.

The following are among the most prominent faults in the construction of the older barracks now existing. Where barracks have been built in forts, they have necessarily been low, bomb-proof, unventilated buildings, as were the Town Barracks in Bombay, the Old Barracks in the Fort of Bellary, in Fort William, &c. &c. Numerous barracks—as Gwalior; the Royal Barracks, Fort William; the Barracks on Mount Aboo; at Bhooj; and elsewhere—have been erected gable-end on to the prevailing breeze. Other barracks have been built under the shadow, and within the effects of radiation from black rocks; in the vicinity of burial-grounds, as at Dinapore; or in damp and malarious localities, as at Aboo; at Colabah, where Sir Charles Napier walked through the sleeping apartments on planks laid in water covering the floor; and at Loodianha, where Dr. Dempster reports that the barracks " wanted in nothing except the means of preventing the monsoon water, flooding the level surface around, from entering the verandahs."

Having fixed upon any spot, high, clear, and dry, and, therefore, apparently eligible as the site for barracks, consideration must next be paid to the subject of drainage; and where from the nature of the soil or surface this cannot be thoroughly effected, there the erection of barracks is forbidden. Clay, for instance, is highly retentive of moisture, and where this material forms the sub-soil, it will keep the ground, and indeed the air over large districts always more or less damp, and is thus unfitted for the location of barracks. Soils which extend to a considerable depth in gravel and sand, with perhaps, a foundation below of marl, and which are so far self-draining, are the best possible sites for barracks and hospitals. Of course, if an elevated and dry spot be chosen, valleys, muddy and marshy ground, will be avoided; and should any unavoidable nullah or ravine exist in the locality, it must be filled up and converted into a surface-drain, if it may not be altogether disposed of. It cannot be too urgently insisted upon, that ravines and nullahs are certain to become the depositories of filth and débris, and hence will not fail to prove injurious in the locality of human residences.

It may seem superfluous to state that a barrack or hospital should not be built over an old grave-yard, or on or near ground charged with organic matter; such, however, has been done in more than one instance.

It should be a *sine quâ non* in Indian barrack and hospital building, that such erections *front* the prevailing breeze. *If the locality will not admit of this being effected, such position ought to be abandoned.* During the Indian hot weather, a building in which air is motionless is oppressive both by day and night. Move, however, several feet into the influence of the breeze, and air in motion produces a sensation of coolness; retire in the latter, and freedom from mosquitoes and sleep will be obtained. Attempt to slumber in Indian still air, and restlessness, debility, and proneness to disease are the results. Again, during the hot weather, the benefit of *tatties* cannot be obtained in buildings where the windows and doors front that quarter from which the breeze does *not* come. I repeat, all Indian buildings ought to be exposed to the breeze which prevails during nine months in the year; and the violence of that breeze should be capable of being moderated by glass windows, doors, and venetians.

If this desiderative is to be carried out, the shape of the barracks does not allow of much choice. The square, as has been proved in other instances besides that of the town barracks in Bombay, is highly objectionable. Barracks *en echellon,* or on the pavilion system, alone fulfil all requirements of perfect ventilation, and such buildings only should be erected. If the military objection be advanced, that this arrangement extends the cantonment to large dimensions, I would reply, that the choice lies between so doing, or otherwise crowding the men, and rendering them liable to all that disease, or tendency to disease, which arises from contaminated atmosphere.

The experience of all ages concurs in the fact that human habitations, particularly in tropical and semi-tropical countries, are healthy in comparison with their height from the surface of the earth (see Chapter III, on Malaria); and it is also admitted that miasm is more powerful on the human system during the hours of sleep or night, than at any other period. Hence the desirability that all barracks in India should be constructed with upper sleeping apartments, the lower being used as reading, dining, or sitting rooms.

The system of building houses and barracks on raised platforms, so common in India, must be but one step less injurious than erecting them on the surface of the ground, as in too many instances the "chaboutra" is formed of rubbish and débris. A striking instance of this was latterly presented in the barracks newly fitted up at Lucknow for a Queen's regiment, where the elements of disease, in the shape of stable litter, were concealed beneath the surface to such an extent as to produce a large amount of sickness, thereby rendering inquiry, investigation, and removal of the nuisance imperative.

For much the same reason, store-houses should never be permitted to exist beneath barracks, becoming, as they can hardly fail to do, receptacles of damp and vitiated air, which, of course, finds its way to the apartments above.

There are, however, other forcible arguments in favour of upper-roomed barracks. The men, during the day, have two roofs over head instead of one, and thus better protection from the sun. The sleeping apartment becomes thoroughly aired and ventilated, which cannot take place if it be occupied day and night. The upper story is cooler during the hot weather than a lower stratum of air would be; any breeze which may be present is felt above when it may not be below; and, in consequence of the greater breeze, and because, like malaria, musquitoes love the ground, men sleeping in upper stories are infinitely more likely to obtain refreshing rest, than if exposed to the greater intensity of disturbing agents which exist below. *Not to sleep at night in India, is to prepare the system for disease, especially malarious disease.* Had all soldiers upper sleeping apartments, we should, probably, have never heard of soldiers taking their "dram," or even intoxicating themselves to procure sleep amongst heat and musquitoes; and thus another source, alcoholic degeneration, and its aid in the preparation of the system for the ravages of disease, would be removed.

The walls of barracks should be constructed of thin stone masonry. Massive walls absorb during the day, and radiate during the night a large amount of caloric, and hence it is advisable, that *strength only* should be considered; the idea that thick and solid walls, mostly composed of mud, render interiors cool being both theoretically and practically erroneous.

Unburnt bricks cemented together by mud—a frequent material of building in India—should never be allowed in the construction of barracks and hospitals. During the monsoon this mud absorbs and retains moisture, and renders the habitation damp long after the surrounding atmosphere is dry. Moreover, when any accidental crack occurs, it becomes the abode of numerous specimens of the insect tribe, who, perhaps, die and decay in the interstices. Stones should be used for the walls of hospitals and barracks, and if stones are not available, fire-burnt bricks. In most Indian stations, however, sandstone, trap, or other rock formation may be easily obtained.

The masonry, instead of being plastered together with mud, should be joined by lime mortar; and, although we are told by Liebig,[1] that houses are unhealthy when first inhabited, in consequence of the lime of the dry hydrate of the mortar combining with the carbonic acid supplied by the lungs and skin, and parting with and setting free as moisture

[1] 'Lumleian Letters,' p. 340.

the 24 per cent. of water chemically combined with it, still, in the heated atmosphere of this country, the above process would soon cease, probably before the building could be in other respects ready for habitation.

All Indian barracks and hospitals which have not upper sleeping apartments should be supplied with a double roof, which, even if collectively thinner than a single covering, will, from the stratum of air between the two layers being a non-conducting medium, prove cooler for those who reside underneath. Although the inner roof may be simply composed of mats or canvass, the air passing horizontally between this and the roof of tiles above, through apertures left for the purposes, will, as Mr. Jeffreys[1] observes, be " an ever watchful corrector of heat." A good form of roof would be that composed of a succession of earthern pots as used in Syria.

Cow-dung or beaten earth floors cannot be too strongly condemned. They are always out of repair, absorb moisture during the monsoon, and emit clouds of dust in the hot weather. Moreover, the periodical application of cow-dung, as layer after layer is smeared on, not only causes dampness for the time being, but adds to the collection of matter which is gradually decaying, and, therefore, constantly emitting a diluted malaria. Dr. Arnott[2] writes—" The barrack floor should always be flagged. No material but stone will stand the thick, iron-heeled shoe of the soldier *mohrusee* and *chunam* are soon broken up, and quickly get out of repair ; and to keep such a floor neat and comfortable is quite impossible."

Greenstone granite, glazed encaustic tiles, slates, or wood are the materials for flooring ; the latter, however, would be the most expensive. Cow-dung, earth, chunam, soorkie, brick, asphalt, and dammered floors, are unsuitable for barrack floors, as they retain moisture, rapidly wear out, and

[1] ' The British Soldier in India.'
[2] ' History of the 1st Bombay Fusiliers.'

the dust of their destruction has a tendency to irritate the eyes, and induce ophthalmia.

Instead of the inner walls being whitewashed, they should be coloured a slightly amber or pink tint. The constant glare of whitewash is painful to the eyes in a well-lighted apartment, and a very small amount of colouring obviates this. Care should be taken that the wall is well smoothed with fine chunam previously to the application of the lime-wash, otherwise organic matter will collect in the cracks, and there decompose. Walls should invariably be scraped before the periodical application of lime-wash, which certainly is a cheaper, and, probably, better method of treating Indian interiors than the application of a "pure white, polished, non-absorbent cement" alone, which has been recommended.

Inner and outer verandahs should encircle all buildings to which an upstairs apartment is *not* attached. Each verandah should be from ten to twelve feet broad, the inner of which may be used as dining, reading room, &c. Where upper sleeping rooms are added, one verandah may be dispensed with, the remaining one being built somewhat broader. It has been recommended to make the roof of the outer verandah flat, as a terrace for air and exercise; but with upper sleeping apartments this could scarcely be requisite.

The absence of light directly tends to lower the vital forces, and render torpid the secreting and excreting processes, and thus predisposes to the inroads of zymotic disease. Moreover the deprivation of sunlight has a great influence in inducing scurvy—a disease which will be shown is endemic in extensive districts throughout India (see Chap. XX, on "Scurvy"). Hence, the windows, while capable of being closed by venetians or "chicks," should be spacious and numerous, facing each other on opposite sides; and care must be taken that the verandah roofs are not brought sufficiently low to darken the interior.

A barrack or hospital built according to this sketch would

require little in the way of ventilation. In India, where doors and windows are so commonly kept open, it is not difficult to secure thorough perflation, provided the residence is situated favorably with reference to the prevailing breeze; *if a building be not so situated, no scheme of ventilation will fulfil the desired purpose.* Hence, natural ventilation is alone to be kept in view in Indian buildings. In a barrack or hospital situated and constructed as described, I would simply recommend ventilation in the roof, and apertures communicating with the space between the two roofs, and with the upper stratum of air in the building, as aids to natural ventilation.

Want of ventilation, insufficient space, or overcrowding are now known not only to induce deviations of health, from the most trivial to the most deadly, but also, without *apparently* being the cause of disease, to aid that deterioration of the system which is always progressing from the heat and malaria of India, and which results in cachexia loci, splenic leucocythæmia, and the enlarged livers and spleens which follow in due course.

Dr. Chevers observes :—" While in temperate climates the crowding together of human beings in small, unclean, and ill-ventilated buildings generates phthisis, diarrhœa, typhus, and puerperal fever, in India the like causes have usually the effect of adding tenfold to the prevalence and destructiveness of the ordinary diseases of the country." Thus, common malarious fever assumes the remittent or typhoid type ; hepatitis, dysentery, and diarrhœa are developed in their worst form ; cholera and heat asphyxia are known by experience to occur ; and fatty degeneration, with its insidious changes, aids malaria, heat, syphilis, alcohol, cachexia loci, splenic leucocythæmia, and the scrofulous taint, in the induction of the more rapidly destructive diseases named above.

It is also well known that a tendency to mental disease is induced or aggravated by the continued imbibition of air

rendered impure by overcrowding. This has been noticed by
Dr. Jarvis in America, particularly in the state of Massa-
chusetts; in England by Dr. Noble, in his ' Psychological
Medicine;' and also has been brought prominently forward
in a late ' Report of the Commissioners of Lunacy in England
and Wales,' from which it appears that mental disease is
forty times more prevalent among the poor, who reside in
crowded localities, than amongst the upper classes, who have
clearer air and more space to breathe in !

It is a peculiar circumstance that small detachments, such
as companies of artillery, enjoy remarkable immunity from
mortal disease. This is explained by the facts that they are
generally well lodged, are seldom or never overcrowded, often
occupying barracks that were designed for larger bodies of
men, while full regiments of infantry are frequently, from
necessity, placed in more disadvantageous circumstances
with regard to air and space. Other causes, however, will
probably exert an influence over the health of the latter; as a
larger body of men attach around themselves, on a more
extensive scale than a smaller number, "those fertile causes
of intemperance, vice, and disease, which accumulate in
every large military station."

Dr. Ewart[1] remarks:—" Evils do not always arise from
overcrowding, but sometimes from want of ventilation in a
large space;" and it must be borne in mind that narrow,
high rooms will not prevent the deleterious effects of contaminated atmosphere. Beds cannot be placed near together,
in the most spacious and lofty apartment, without exposing
their occupants to a vitiated atmosphere, and obliging each
to inhale the carbonic acid just exhaled by his neighbour.
Hence, barracks must be spacious enough to allow each man
seventy feet of superficial space, and as much as 1200 feet of
cubic space, as the *minimum* in tropical climates.

---

[1] Ewart, ' Sanitary. Condition of Indian Gaols.'

Dr. Walker,[1] in his ' Report on Epidemic Typhoid Fever occurring in 1860, in the Agra Gaol,' remarks—" We are too apt to apply the tape-line, and measure out a certain number of cubic feet of space as capable of supplying a human being with air to fill his lungs and keep him in health," and insists upon the importance of relieving the atmosphere of the breathing and exhalations which arise from the *continued* occupancy of any one space by bodies of men.   This, as already remarked, forms a cogent argument in favour of upper-roomed barracks, and also for the occasional removal of the occupants of Indian prisons into tents or other buildings.

A frequent and powerful source of vitiation of the atmosphere in barracks arises from decaying teeth, which a large proportion of the men, in tropical climates especially, will be found to possess. The effluvium arising from a decayed tooth is probably sufficiently well known, and when many such exist in the persons of soldiers sleeping in a common room, the contamination of the atmosphere caused therefrom must be equally obvious.   This affords another reason for the employment of a soldiers' dentist, as recommended in Chapter XVIII, on Diet.

All corner-rooms, made by partitioning verandahs or otherwise, are most objectionable, being necessarily small, hot, and unventilated, and denying thorough perflation of parts behind.

The method of cooling Indian buildings is a subject of great importance, not only to the comfort but also to the health of the occupants.

Mr. Jeffreys[2] describes a plan of ventilating and cooling barracks by the subterranean absorption of heat, the process consisting of conveying air through a multitude of wells sunk

---

[1] ' Government Report on Epidemic Fever in the Gaols of the North-west Provinces,' 1861.

[2] Jeffreys, ' The British Army in India,' p. 132.

in front of the barrack, one shaft of which opens in the interior of the building. This, however, would only thoroughly take place during the prevalence of high winds; and it is not improbable that air so passed through the ground would become saturated with damp and malaria, as, indeed, is reported to have been the case when the system was tried, thirty-five years since, at the Cawnpore European Hospital.

Until some better means for cooling large buildings can be devised, recourse must be had to the old system of tatties, punkahs, and thermantidotes, which, notwithstanding the inconveniences attaching to their use, are unquestionably more conducive to health than allowing the uninterrupted play of the hot and arid wind upon the skin, or, in other words, allowing the surface of the body to act as a tatty.

With tatties, constant vigilance only suffices to keep them from becoming dry; and, from partial evaporation or greater force of wind, the temperature of a room so cooled rises and falls many times during the day eight or ten degrees, which is certainly trying to the constitution, and tends to induce rheumatism and bronchial complaints.

If punkahs or thermantidotes only are used, all doors and windows being shut, as is necessary to exclude the hot air, the atmosphere of a room containing many occupants, although kept in motion, would become hot, close, and saturated with effluvia. If ventilating apertures were formed sufficiently large to obviate this, they would be capacious enough to allow the passage of sufficient heated air to negative the effects of the fans; and although it is undoubtedly better that men should breathe hot pure air than foul cooler atmosphere, still a judicious use, both of tatties and punkahs, thermantidotes, or other blowing machines, will secure the advantages of both, with a *minimum* of the injurious effects arising from either. Hence it appears desirable that barracks and hospitals should be supplied both with ventilating fans and tatties in certain proportion, according to the size of the building and force of the wind. The tatties, without presenting surface enough to cause those sudden alternations of temperature spoken of, should have an area sufficient to afford free perflation. The fans would immediately disseminate the air from the tatties equally throughout the apartment; and the small ventilating apertures, previously described, with the ill-fitting doors and windows inseparable from an Indian climate, while sufficient to promote the action of the tatties, would not be large enough to heat the apartment, which would be supplied with pure air through the latter.

For the ventilation of barracks, however, during the period the heat renders it necessary to close doors and windows, a much superior method would be the construction of a large chimney, with an opening on the roof, which would allow of being directed towards the prevailing breeze. This would, in fact, be applying the principle of the wind-sail used on board ship for the ventilation of the vessel. A tatty in the lower opening of this chimney would effectually prevent the entrance of heated air.

In those districts—as in Lower Bengal—where, from the absence of winds, tatties will only act imperfectly, the plan of extracting air at one end of the building (as employed in certain factories in Europe) might be introduced, and the atmosphere suffered to pass in at the other extremity of the range through tatties erected for that purpose.

Care must be taken that a sufficient number of fans, punkahs, &c., are used to keep the *air in motion* throughout the *whole space* of the apartment; otherwise, those who happen to be in the direction of the column of air will receive all the benefit—their neighbours none.

I am not aware that steam-power has been used in India to work cooling apparatus, but no doubt can exist that machinery might easily be adopted, which would move both punkahs and thermantidotes, or other blowing machines, and water the tatties at the same time. Such machinery would, I imagine, prove less expensive than manual labour.

Of late the ventilation and warming of large buildings has received considerable attention, not only in England, but on the continent of Europe also. Of the various systems proposed, that of Van Hecke would appear to be the most successful. "This system," says Dr. Pettenkofer, of Munich, in his recently published ' Remarks on Warming and Ventilation,' " has completely upset all our ventilating traditions, and is now employed in the Chamber of Representatives, at the Hague, in the Hospital Necker, at Paris, and at other places. The method appears extremely simple, and has been latterly brought before the Metropolitan Association of the Medical Officers of Health, by Mr. Phipson, C.E. Fresh air is propelled along a channel, by means of a peculiarly shaped fan, into an air chamber, containing a warming apparatus, where it is warmed and moistened, and whence it is distributed over the whole building. An anemometer and dynamometer placed before the fan indicate at any moment the amount of air supplied. The *minimum* is stated to be 2200 cubic feet per hour per patient, supplied without draught.

If, as undoubtedly might easily be effected, a cooling chamber were substituted for a warming space, such an apparatus would be invaluable as an adjunct to hospitals, barracks, public buildings, or even private houses in India.

The necessity of out-houses being placed to leeward of inhabited dwellings, has already been dwelt upon in the chapter on " Cantonments and Stations."

All barrack latrines should be placed to leeward of the habitations, or at all events in such positions that the prevailing breeze cannot pass from their quarter towards the barracks. The seats should always be furnished with lids, the aperture should consist of a whole circle, instead of a semi-circle, and the seat should not be so high as to prevent a moderately tall man placing his feet on the ground when using the convenience. These buildings should be at least thirty or forty yards from the barracks; and as the great majority of individuals use them at morning or evening, the distance would not lead to exposure to the sun. They should be kept clean by bi-daily manual labour. (See chapter on " Conservancy.")

Both bath-rooms and urinals should, I think, be built under one roof (with a partition wall), in order that the water from the former may be used daily, or as occasion may require, to flush and cleanse the latter.

Urinals, although kept in a very cleanly state, will at times emit an offensive and unhealthy effluvia, which, if the building be placed too near the barrack, will, in calm weather, and notwithstanding a leeward locality, be recognisable in the latter building. If the urinals are placed too far off, the men will hesitate to use them, and micturate on the ground. In order to meet these objections, the urinal should be about thirty feet from the barracks, with a covered communication, raised, if such be necessary from the condition of the surface, several feet from the level of the ground. The urinal should be a long building, with compartments and trough for the reception of fluid, and raised four or five feet from the earth. In front, steps should lead into the interior; while behind, doors should open into the cavity between the under surface of the floor of the urinal and the ground. In this cavity should be placed iron receptacles for the collection of the urine, which should flow from the troughs by a communicating pipe. The floor of the urinal should also slope, in order to prevent lodgment of accidentally spilled fluid, and also communicate by an additional pipe with the iron chest below. This chest should be emptied and cleansed night and morning by the conservancy establishment, and be furnished with handles, to allow of its being carried away for that purpose.

The bath-room should also be raised from the ground, and be divided into compartments, with a trough for the reception of "ghindees," rather wider and at a higher elevation than the trough of the urinal. This would enable the latter to be flooded from the former, as requisite, through the medium of a pipe in the division wall. Tubs may also be placed in the bath-room.

The exit of the water has, however, to be provided for; and if the surface of the ground is such as to prevent drainage taking place by gravitation, a well should be sunk underneath the bath-room, into which the water would be conveyed. In most soils, at a certain depth, water is found, and it cannot be argued that an addition to this moisture, caused by the bath-room water, would make the locality more unhealthy. The simple fluid certainly would not, and the small amount of dirt from the bodies of the men, or the salts from the soap used, could not be sufficient to induce disease, all entrance of débris into the well being effectually stopped by a grating convex upwards over the orifice. This plan has been recommended by Major C. B. Young, Chief Engineer in Bengal, and, I believe, actually practised for the reception of urine! Masses of urine, however conveyed into a well, must quickly develop a pestiferous fluid mass, and on sanitary grounds is to be strongly condemned. With water, however, the matter is different; and I believe the suggestion of this method of disposing of bath-room water was first made by myself, in a recent recommendation with regard to the Mount Aboo Barracks. Whatever objections may be taken to the plan, it must certainly be a better one than allowing the ground to become wet, sodden, drying, and malarious from the water flowing on the surface. On the other hand, where capacities for drainage exist, the latter should be had recourse to, and, until a well or other method of disposal is established, manual labour should be employed for the removal of the water.

There are many objections to the disposal of urine, ordure, or refuse, by conveyance into running streams, as such procedure pollutes the water, and frequently induces disease, or tendency to disease. (See chapters on "Cholera," "Water," and "Encampments.")

CHAPTER XIV.

ON HOSPITALS.

Dr. Moseley long since wrote—"It is a solecism in economy to have a bad hospital," and his experience at Castille Fort where "in three months more value in men was lost, from the miseries of the hospital, than would have been adequate to the expense of erecting a proper one for all the troops," is the experience of every medical man who has seen human beings massed and crowded in small, ill-constructed, ill-adapted, or unventilated buildings, especially when such individuals have been emitting the emanations of diseased action.

As the writings of Cæsar and Xenophon make no mention of hospitals for the reception of the sick and wounded, it has been asserted that the records of ancient times are silent regarding such establishments. This, however, is not the case, as the organization of the "valetudinarium" in the Roman camp, is described by Hyginus Gromaticus, and the same is referred to under the term "contubernales" by

Vegetius. We have also authentic data[1] that four *medici* were attached to each cohort, and Velleius Paterculus, in his account of the expedition into Germany, describes the provision of physicians, and of other requisites for the health of the army, as in such profusion, that "only home and domestics were wanting."

From the eminent discipline of the Roman army, from the physical force which enabled Cæsar to write "Veni, vidi, vici," it is unquestionable that the so-called civil departments— on which, as Sir Ranald Martin truly observes, "not only the efficiency, but the very existence of armies depend "— must have existed in excellent executive condition, as an intrinsic portion of the Roman system.

Be this, however, as it may, it is none the less a fact that our systematic treatises on medicine are silent on the important subject of hospital construction. Only scattered writings referring to the matter, and these from the pens of military surgeons, are to be found. These commence at the recent date of 1764, with Pringle's work ' On Diseases of the Armies,' wherein separate regimental infirmaries, and isolation of the sick, is advocated. This was followed by the ' Medical and Economical Observations of Brocklesby,' published 1758, and the latter by the works of Sir Gilbert Blane, Iberti, Mr. Henderson, and Sir W. Blizard, all of which refer more or less to hospital construction.

During the first fifty years of the present century, with the exception of a paper by Sir George Ballingall in the ' Cyclopædia of Practical Surgery,' nothing appears to have been published regarding the construction of hospitals until, in 1856, Mr. Robertson read before the Manchester Statistical Society a paper on the subject. Since that period the matter has been much discussed in the columns of the ' Lancet,' ' The Builder,' ' British and Foreign Medico-Chirurgical Review,' and other periodicals; the mortality in

[1] Orellius et Henzen, ' Inscript. Latin Select. Collectio.'

the hospitals at Scutari, in 1854-55, having apparently revived the opinions and fears previously expressed regarding ill-adapted hospitals by Pringle, Brocklesby, Blane, and Blizard.

A recent writer observes:—"It is with the greatest difficulty that any important principle can be made to attract the attention of the public. Doctrines must be embodied in some material or substantial form, before they make any impression on the general community." So far as the construction of hospitals is concerned, this is most true; such impression, however, was made during the Crimean war, and its fruits were apparent in the exertions of Miss Nightingale, Lord Herbert, Dr. Sutherland, and some others.

It now demands that the experience of the last decade, and the warning voices which have arisen, shall not be passed over, as were the writings of Pringle, Howard, Hales, Blane, and others of the last century. If unhappily such be the case, if large hospitals are built defiantly as regards sanitary requirements, if crowding again takes place, if hygiene be not strictly attended to in every respect, then the certain results to which we may look forward will be, a repetition of the Crimean hospital mortality, a second edition of the mortal sickness of the Hôtel Dieu, in 1771, and the experiences of Howard, Pringle, Lind, Blane, and the authors of the last century, again displayed in the next generation.

It has been remarked that the practice of physicians, whether good or bad, according to orthodox medical views, does not materially influence the ultimate mortality of an hospital. Of this I was afforded a striking proof during the three years I was resident surgeon at the Queen's Hospital. To this institution three physicians were attached, of whom one treated his patients with stimulants scarcely in less quantity than the late Dr. Todd; a second had a decided preference for the Hamiltonian system of blue pill at night and black draught in the morning; the third treated his patients without any peculiar bias. On examining the records,

I found there was very little difference between the mortality
and cured among the patients of the three doctors, who I
trust will pardon me this mention of their practice.

It has now become an established fact, that a large town
is usually less favourable to health than a small one, and the
country is even more conducive to efficient sanitation than
the latter. Miss Nightingale remarks—"If the recovery of
the sick be the object of hospitals, they will not be built in
towns;" and although, for obvious reasons—as the reception
of acute medical and surgical cases—hospitals can never be
wholly located in the country, still, with military hospitals
the conditions are different; and henceforth sick soldiers
may be removed from London to the purer air of Netley;
from Bombay to the bracing atmosphere of the Western
Ghauts; from Calcutta to the pure ocean air of the Sand-
heads; from Madras to the Neilgherries.

Pringle long since recognised the evil effects of massing
together great bodies of sick in general hospitals, thereby
generating a "hospital atmosphere;" hence his recommendation of separate regimental infirmaries—the germ of the
regimental hospital system of the present day. "Hospitals,"
wrote Pringle, "are among the chief causes of mortality in
armies, on account of the bad air, and other inconveniences."
In every campaign he records the fatal effects of hospital
fever, and lays it down as a rule that the more air that can
be let into a hospital the less danger to sick and attendants.
Separation of the sick in separate regimental infirmaries,
and space sufficient in those infirmaries to make a person
unacquainted with bad air imagine there was room to take in
double or triple the number, were the means of escape from
the diseases generated in hospitals, which this author so
urgently pointed out.

To find that construction which will accommodate the
greatest number of patients upon a given area, with the
greatest facilities for economy, administration, and recovery in

the shortest possible time, is still a desiderative as regards large general hospitals. In fact, with the knowledge that pyæmia, erysipelas, surgical fever, irritative fever, &c., are still frequently epidemic in our best London and provincial hospitals, we may still demand with Pouteau, " Are hospitals more pernicious than useful to humanity ?"

The subject of the construction of large general hospitals is too extensive to be entered upon here ; and the following remarks are intended as chiefly applicable to Indian regimental hospitals in the " Mofussil."

The first point to be attended to is, undoubtedly, the selection of the site ; but in this there is not much choice. Where the locality of a cantonment has been well selected, and the position of the barracks chosen with due attention to sanitary requirements, there will be little difficulty in arranging for the hospital, which should be placed as near the former as the requirements of healthy locality, ventilation, and space will admit ; *but the sanitary demands of this building above all others must not be sacrificed to beauty and uniformity of architectural design.*

The general plan which embodies the only sound principle of hospital construction is that which is known as the " pavilion system," that is, the separating or breaking up of the hospital into a number of distinct erections, capable of containing about fifty sick each. The simplest plan is to build such a hospital in a straight line, fronting the prevailing breeze, and leaving all angles open to permit thorough perflation.

Pages might be written, and hundreds of instances adduced, of the danger of placing surgical and medical patients in the same apartment. Compound fractures, or other severe surgical cases, cannot be expected to progress favourably in the immediate neighbourhood of bowel complaints and fevers, surrounded by air contaminated by the exhalations from the lungs, bodies, and fæces of such patients.

Neither can medical cases be favourably influenced when exposed to the effluvia from large, unhealthy, or gangrenous sores.   Dr. Chevers[1] writes :—"I am firmly convinced that, sanitate as much as we may, we shall never be quit of the danger of surgical fever, until we make proper arrangements for the segregation of our surgical patients."   I am quite of this opinion; and moreover believe that, especially in cities and malarious countries, a medical case is quite as dangerous a neighbour to a surgical case as the latter is to the former.

Separation was an arrangement insisted upon by Howard as a *sine quá non* in hospital management, and has since been advocated by Blane, Pringle, Sir J. M'Gregor, Dr. Simpson, and others. Dr. Simpson observes:—"Hospitals get deteriorated by long use. Old surgical wards and old surgical hospitals seldom offer such good returns from practice as newer hospitals and newer wards."[2]

Thus, an Indian hospital should consist of at least three detached wards : a surgical, a medical, and a woman's ward. But it is also desirable that a fourth or convalescent ward should be established; and not only this, but a fifth, for the reception of all contagious diseases; as scabies, cholera, smallpox, &c.

As, however, surgical cases are not so numerous in Indian practice (times of warfare of course excepted), the surgical ward may be considerably smaller than that intended for medical cases.

I also think that in all large stations a building might be appointed for the reception of contagious cases from every corps belonging to the brigade, and that the sick might be there treated by their own medical attendants.

The advantage of a convalescent ward will not be questioned. The sick unto death and those rejoicing in returning health should not sleep and live together. The noise of the latter

---

[1] Chevers, 'Ind. An. Med. Sci.,' vol. xiii.
[2] Simpson, "On Diseases of Women." 'Med. Times,' May, 1859.

annoys the former, and the former in their turn depress the spirits, and hence retard the recovery of the convalescing. In fact, the Queen's Regulations of 1859 demand that in all hospitals, when practicable, wards should be set apart for convalescents.

The reasons given for recommending upper-roomed barracks are applicable with still more force to hospitals. The latter should always have upper apartments, and the lower might be used as convalescent, day, reading, or sitting rooms. Next to upper stories, the plan of elevating hospitals on arches is most desirable.

The roofs should be double, and the walls faced internally with polished cement, which, like lime-wash, does not present a rough surface for the lodgment of organic impurities, always arising from the bodies and lungs of the sick. The flooring should be composed of tiles, and no cow-dung or other material capable of decomposition should ever be placed underneath or upon the lower structure.

The most convenient location of the hospital attendants, as the apothecary, serjeant, and other assistants, is a question of some little difficulty. When hospitals are constructed with upper rooms, it may be allowable to convert some part of the lower story into apartments for the subordinate officers. If, however, upper rooms are not provided, these essential attendants should be accommodated in detached buildings; all corner rooms, divisions and partitions of the main structure tending to deny thorough atmospheric perflation, and therefore, if possible, are to be avoided. This, however, as will be immediately referred to cannot always be effected in hospitals.

A hospital ought to have none of its offices in duplicate. It should have one kitchen, so situated that the diet can be easily carried to any ward on the shortest possible notice. Of course neither kitchen, sewer, nor drain should be allowed to exist *under* any part of the building. Bath-rooms, closets, and latrines should be placed to leeward, but communicating with the hospital, in which a corner room must be allowed to be used by the seriously sick only, as the latter. This room should be supplied with trapped portable pans, which, with others of the same description necessitated at the bed-side, should be immediately removed by the sweepers. The use of any description of pipe or drain is liable to lead to deposit and unhealthy exhalations, and is inadmissible in an Indian hospital.

Dr. Alison remarks that when fever, erysipelas, or gangrene spreads in a hospital, or originates in its wards, such an event means that the laws of nature, as to the abundance of fresh air, are not being carried out. Hence the ventilation of a hospital demands even more care than is required for the perflation of barracks, as deleterious emanations are given off in increased abundance and with increased rapidity from the sick, as compared with those from the healthy. Thus the air of a ward is liable to become vitiated in a much smaller space of time, when filled with sick, than if it contained the same number of healthy individuals; and, moreover, certain diseases, as fevers and surgical cases, contaminate the air more rapidly than others.

The same means, however, which have been recommended for the ventilation of barracks, combined with every aid from natural ventilation, is the only plan which can be adopted in India with any great hopes of success. Artificial ventilation has been sufficiently practised in European hospitals, but has not yielded results so satisfactory as to counterbalance the enormous expense attending it, excepting on the Van Hecke principle, previously referred to. Newly invented " self-acting" punkahs are now being experimented on in Bengal, the value of which, however, in large buildings, as hospitals, has, I believe, yet to be demonstrated.

Every Indian medical officer will admit the great existing want of a hospital to which sick officers might be taken ; and such should be built near every Indian cantonment. An erection on the plan of a traveller's bungalow would suffice for the purpose. A European nurse should be retained, with a sufficient number of native servants, and the whole placed under the control of the station staff medical officer.

The advantages of such an establishment would be manifold. At the present time, the sick bachelor officer lies in his bungalow, dependent on the kindness of his comrades, and on the oftentimes lazy and unwilling attendance of his native servants; moreover, as the young officer can seldom afford to keep a good cook, he is dependent on the mess for his sick wants, and is frequently vilely supplied. Should he become delirious or helpless, he must either be left to the care of natives, or become a tax on the kindness of his brother officers, who, to their credit be it said, are ever ready to feed, tend, wash, to sit up with—in fact, to act as nurses. This, however, should not be; a man cannot perform the office of nurse; it is a woman's province; and *thousands and thousands of officer's sick beds have demanded, and are still demanding, that care which woman alone can give.* Many an officer who goes home sick, would not have been obliged so to do had he obtained good nursing during his illness in this country, and thus this expense and loss of service would have been saved to the State. Whether, however, the establishment of officers' hospitals and nurses would prove economical, or the reverse, there can be no question that *humanity* calls for their presence. I believe nurses have been organized in Calcutta, for service there, and in some of the larger Bengal stations. The sooner such is the case all over India the better; and I commend this subject to any lady ambitious of following in the glorious paths of Miss Nightingale.

CHAPTER XV.

ON CONSERVANCY.

Dangers of Neglected Conservancy: Remarks of Board of Health; Application to Indian Stations—System of Conservancy pursued in India—Condition of Indian Cities—Séparateur System recommended—Composition of Sewage Masses—Effects of Inspired Sewage Air—Heat a Safeguard against Disease—Bounty of Providence—Indian Conservancy Establishments—Regulations concerning Sanitary Officer—Disposal of Ordure—Bazaars—Public Latrines—Sanitation to be Compulsory in Indian Villages near Cantonments—Deodorising Agents; Charcoal.

FOREGOING chapters prove that it has become an established principle that neglected conservancy will almost inevitably give rise to typhoid fever, typhus, diarrhœa, dysentery, cachexia, or some other deterioration of health, according to the nature of the climate, the diet, mode of health, and other concomitant circumstances.

It has been shown in previous remarks that epidemic pestilences are the direct consequences of certain meteorological conditions combined with a local cause, such causes being filth, moisture, stagnant, and soiled air, and especially the emanation arising from decomposition. Indeed, as will be afterwards referred to (Chapter XVII, "On Cholera"), there is strong reason for supposing that the cholera germ itself is nothing else than the choleraic fæces in a state of decomposition. We also know that dysenteric dejections in the same condition will multiply dysentery.

The General Board of Health, in one of their minutes, remark :—" The habits of a people with respect to cleanliness and more especially with respect to their care to protect their habitations from pollution by excrementitious matter, are clear indications of their progress in civilisation. Archdeacon Paley was accustomed to direct the particular attention of travellers in foreign countries to the mode in which people dealt with their excrete, stating that, from this fact, 'a greater insight might be gained into their habits of cleanliness, decency, self-respect, and industry, and in general into their moral and social condition, than from facts of any other class."

And again, in the minute dated 1852, it is laid down as established :—" All offensive smells from the decomposition of animal and vegetable matter indicate the generation and presence of the causes of insalubrity and of preventible disease, at the same time, that they prove defective local administration."

That this is the case in numerous Indian stations, is painfully evident from what has been stated of Simla, Ootocamund, Mount Aboo, from what Dr. Balfour writes of Delhi —" That the rocky range called the Hindoo Rao is made a temple of Cloaca, so that when heavy rains fall, polluted streams run down as far as the Sepoy's lines, where the slope stops, and where nearly all the moisture is absorbed"—and from very many other reports which might be quoted.

The system of conservancy pursued in India is that known as the dry system ; and indeed, in our Indian cantonments, none other, excepting the objectionable one of cesspools, could be employed. Sanitarians, however, in all civilised countries, have been long acquainted with the pernicious nature of the effluvia exhaled from such receptacles, and hence have done their utmost to convince the public of the deleteriousness of sewage matter, when allowed to accumulate and undergo chemical decomposition in or near the habitations of human beings.

Any other plan excepting the two mentioned above would

be most difficult and costly to establish and carry out in an Indian cantonment; the distance between the different residencies rendering a system of drains too expensive, and the want of a supply of water in most stations forbidding the possibility of carrying off refuse by such means, even if no other objections to this mode of disposing of débris existed.

It is chiefly due to the exertions of sanitary reformers, as Martin, Mackinnon, Hathaway, Chevers, Lownds, and others, that the dry system has obtained in India instead of cesspools.

In the thickly populated districts of large Indian towns, as the presidency cities, drainage is employed, but of such a character as would, in the words of a recent author,[1] "disgrace a nation just emerging from barbarism." A very short walk through any Indian city is quite sufficient to demonstrate that the germs of disease pervade every locality; and the fruits of this are evident in the mortality which exists in such cities, and in the almost constant prevalence of cholera, either in a sporadic or epidemic form.

"In point of sanitary philosophy," writes the editor of the 'Bombay Times,' "it is hard to say which of the presidency cities is worst off. According to current testimony, there is not one of them that is not justly described as a 'stinking hole.' A few days' absence from Bombay, and return, is all that is required to make one conscious of the filthy atmosphere in which we live."

From a recent report, it appears that the system now generally adopted in Paris for the removal of sold and liquid excreta, is that of movable *séparateurs*. The *séparateur* is an iron box, open at the top, and the sides of which are perforated with numerous holes, the purpose of which is to retain all solid matters, and to allow all liquids to percolate into the sewers, by which they are conveyed away. It is so fixed as to be easily removed and reinstated, and it is said

[1] Lownd's 'Sanitary Aspect of Bombay.'

that this process occupies but a few minutes, and gives rise to no annoyance whatever, the full boxes being placed immediately on their removal in an air-tight case for transportation.

This *séparateur* system not only prevents, in a great measure, the pollution of the river, or wherever the drainage is received, but is also free from the serious drawback arising from the existence of an impure subterranean atmosphere of immense extent, which exists in the sewers of all large towns; or if such impure atmosphere does exist, it is but in a less intensity, and may be readily destroyed by substances capable of averting putrefaction—such as mineral acids, creosote, and carbonic acid, or of substances capable of decomposing and destroying putrescible matter, as chloride of lime, &c.

Sewers sufficiently large to convey solid matter must, in time, become depositories of the same, and, of whatsoever material constructed, will allow the permeation of deleterious gases, which eventually find their way into houses, not only through every chink and crevice, but also *through the structure* of the sewer itself.

Hence it appears undoubted that the *séparateur* system is that most adapted to tropical cities, as Bombay and Calcutta, for instance, the solid refuse being taken away, and the liquid matters allowed to flow into the sea or the river, through a system of small drains.

Dr. Letheby, in his report to the Commissioners of Sewers, 1858, states that the chief constituents of sewage masses, are carbonate of lime, common salt, and alkaline sulphates and phosphates. The animal products are spirulina, vibriones, monads, paramecium, and other infusoria. Vegetable growths are oscillatoria and conferva. Gases are sulphuretted hydrogen, a powerful narcotic poison; carbonic acid, also a narcotic; ammonia and its compounds with the two former, and coal gas. The effects of breathing sewage effluvia has already been referred to (Chapter II, "On the Causes of Zymotic Disease").

If such be the case in the colder climate of England, where putrefaction in closed drains and cesspools does not occur with half the rapidity with which it takes place in the tropics; where the atmosphere contains in any given bulk —from its more rarified condition in hot climates—more of the vivifying oxygen; where the vital powers of human beings are not degenerated or prostrated by heat and malaria; where the extremes of dryness and moisture do not exist at different seasons—how much more inimical to human health must the emanations from putrefactive process be in India?

Our great safeguard against this quicker and more extensive diffusion of deleterious gas in India, from neglected conservancy consists, as has already been hinted, in the very heat of the climate itself, where " baggage animals drop down dead on the line of march, but there is no decomposition; they appear to dry up in the sun like mummies."[1]

Did not this scorching take place in the hot weather during four or five months in the year, and to a certain extent in the cold and rainy season also, the neglected conservancy of Indian cities and stations, the daily masses of ordure deposited by the natives, and, indeed, by the Europeans also, *on the surface of the ground*, would quickly induce a pestilence such as the world has never yet seen.

Providence, however, is bountiful, and will not allow human beings to suffer beyond a certain extent. Providence, however, while providing means to establish a certain equilibrium between health and disease, " helps those still more who help themselves;" and sanitary art points out the way to a preventive system, which, with the blessing of the same overruling Power, may ultimately free our Indian possessions from all the mitigable class of diseases.

[1] Lowe, "Med. Hist. of Mad. Sappers and Miners;" 'Madras Quarterly Journal,' No. 1.

In most sanitary cantonments, a conservancy establishment is entertained, consisting of men and carts, and such are either placed under the barrack-master, if there be such a functionary, or otherwise under the commissariat department.

By Her Majesty's recent regulations, however, as before stated, the senior medical officer at every station is, *ex officio*, the sanitary officer, and all medical officers communicate with him on sanitary subjects, as a part of their ordinary duty. The senior medical officer, however, has no power to carry out any sanitary improvements, and, as just pointed out, the conservancy establishment is quite independent of him. Under such circumstances, but little good can be expected to arise. The arrangement is one which cannot work satisfactorily. The sanitary officer should be as supreme in all matters connected with preventive medicine as he is now with respect to curative medicine. Certain regulations, public to all, being laid down for his guidance, "abuse of power" would be impossible.

At present, the *ex officio* sanitary officer performs his duties without extra allowance. If, however, such duties are to be rendered satisfactorily, they demand a larger amount of energy and knowledge of specialities than suffices for the daily routine of the most successful medical practice. Moreover, it may be taken as an axiom, that unpaid labour is generally not satisfactorily performed. Therefore, as there certainly is little honour attaching to such an appointment, and as it is beyond the possibility of doubt that such duties, if well performed, would prove an immense saving to the State, it is not too much to suppose, that the time is not far distant when the sanitary officer will be paid for his work.

All ordure, or other putrefying animal or vegetable matter, should be taken by the conservancy establishment at least two miles, in the opposite direction to the prevalent wind, from camp limits, and there daily buried in trenches or

pits. At some seasons it might be used to fertilise any
cultivated land existing to leeward of the cantonments; but,
as a general rule, it should be covered with earth, after as
much as possible has been consumed. The trenches or pits
need not be more than four or five feet deep; otherwise
combustion will not take place. At this depth, however, all
inflammable material, such as stable litter, &c., would
burn, and the bulk of the mass be sufficiently reduced to
allow of its being covered with several feet of earth.

It is little good carrying refuse a few yards without the
camp limits, and depositing it in heaps, as is frequently done.
The mode prescribed by the Mosaic ritual—a method resulting from an inspired prophet—is the only satisfactory
manner of disposing of excrete in a tropical country.

Dr. Hathaway states the best way of disposing of refuse
matter from gaols is, to have a trench cut at the further end
of the garden, to be kept ready dug, and filled up. Decomposition is effected in about six months. Earth is a capital
deodoriser, as the grave-yards of every city testify, and, if
sufficiently covered, no danger is to be apprehended from this
method of disposal of ordure. Prisoners should be marched
from the gaols to trenches at a considerable distance, each
morning or evening; a plan pursued at the Kurrachee gaol,
when under the superintendence of Major Arthur.

It is, however, the bazaars, and the rear of bazaars attached
to Indian cantonments, where sanitary regulation and conservancy are most required. Any ravine, nullah, or clump
of trees becomes, like the rocky slope of the Hindoo Rao at
Delhi, a temple of Cloacina. This should not be permitted.
Health and decency alike forbid. Public necessaries, both
male and female, should be constructed, to which all should
be made to resort, and these places should be daily cleansed
by the conservancy establishment.

In a former chapter, (Chapter X,) the subject of drains was
mentioned. These, in all cantonments, should be open, and

only used for efflux of rain-water. Moreover, they should be swept from one end of the cantonment to the other by the public sweepers.

In all large towns which are near to cantonments a system of conservancy should be made compulsory on the inhabitants. When certain objects are to be effected, the natives of India show no want of alacrity in combining to effect them. Tanks and temples have been constructed in this way, without the assistance or interference of Government, in accordance with the conceptions of the inhabitants as to what was needful and proper.

Their idea, however, with regard to sanitary matters being very primitive, all large towns in the *vicinity* of military stations should be compelled to adopt the machinery of Act xxvi, of 1850, for enabling the inhabitants of any place of public resort or residence to make better provision for purposes connected with public health and convenience; or otherwise to find their own machinery, as " punchayets," for effecting what might be necessary to be done.

The latter, indeed, would appear to be the easiest mode of dealing with the conservancy of native cities. State what is to be done, and leave the natives to carry it out in their own manner.

In Indian towns and villages no one thinks of combining with his neighbour for anything which is not *immediately* productive, and therefore, it requires authoritative orders before the natives will move in sanitary matters. In all bazaars, villages, or towns in the neighbourhood of cantonments, sanitary measures should be insisted upon, otherwise, from a disregard of the commonest precautions, the luxuriance of vegetation will not be repressed, the tank will not be cleared out, the stagnant water find no outlet, and the diseases and squalor visible in the faces of the natives be conveyed to their European neighbours.

A few shovelfuls of earth would repair the village road or

the narrow embankment which goes right across the plain, and serves for intercommunication during the rainy season, and three trunks of trees with some earth would make a bridge over the deep ditch at the outskirt of the village ; but men and women tramp through the water up to their middle because these simple repairs are not executed, and because the inhabitants roar to the Hindoo Jupiter, instead of getting out of the mud themselves. These evils will only be cured by enlightenment and education ; but in the meantime they should not be endured in the neighbourhood of military cantonments.

In both public and private necessaries—the latter including those attached to barracks and hospitals—deodorising agents should be made use of ; and a few words on this subject cannot be misplaced.

Dr. Letheby[1] states :—" The right knowledge of the action of disinfectants dates from a very recent period. They all act in one of two ways; they either give stability to the organic matter, and so check its tendency to decay, or they operate on the putrid vapours, and destroy their offensive properties." This they do either by fixing the effluvium and forming compounds which are inert—thus breaking up the putrid molecule and changing its nature—or by expediting the process of decay, and hurrying it to the last stage of oxydation.

Those substances which give stability to organic matters are called " anti-septic" or " anti-putrescent." Salt, sugar, vinegar, creosote, and the empyreumatic oils, alum, and the astringent matter of many vegetables, are examples of this class. They are, however, scarcely applicable for purposes of deodorising and disinfecting, although of extensive use in manufactures and arts.

Of the second class of substances, those which may be called deodorisers, or, in other words, those which combine

---

[1] "Sewage and Sewer Gases ;" 'Sanitary Review,' October, 1858.

with the putrid gases and fix them in an involatile form, there are numerous examples. Such are the metallic oxides and their salts, as sulphate and chloride of zinc, the latter forming the patent compound known as Sir William Burnett's Disinfecting Fluid. Acetate and nitrate of lead ; sulphate, muriate, and pyrolignate of iron ; muriate of manganese ; the fixed alkalies, and some salts of lime also act in the same way. Most of these agents unite with the sulphuretted hydrogen and ammonia of sewage matters, and, therefore, destroy the offensive smell, but do not affect the organic vapours. All these, however, are too costly for general and extensive use.

Of the same class some substances act chemically on volatile matters, and form new compounds, which are inert. Of these, which may more probably be designated disinfectants, the most important are chlorine, chloride of lime, hypochlorous, sulphurous, nitrous, and carbolic acid, the latter now recommended by Major Curtis—one of the municipal commissioners—for use in Bombay. The first of these, viz., chlorine and its oxi-compounds operate by extracting hydrogen from putrid vapours, and by decomposing water and setting free oxygen to destroy miasm. Those substances containing chlorine may be described as both deodorising and disinfectant, as they not only act in the manner just stated, but also unite with sulphuretted hydrogen and ammoniacal compounds. Chlorine compounds, however, are costly, and like sulphurous and nitrous acids, when in excess, are powerful irritants, and even dangerous of themselves.

The last class of disinfectants are those which expedite decay, and cause putrid matters to combine with oxygen. Of this class there are two divisions, namely, those which act chemically and supply oxygen of themselves to offensive compounds, and those which merely facilitate oxidation by their physical properties.

The manganates and per-manganates of potash are the

best examples of the first class. They contain a large
amount of oxygen, which they give freely up to putrid
organic matter, and so destroy it by a slow process of oxidation or burning, which is called eremacausis. Dr. Angus
Smith's experiments on foul air with permanganate of potash
have been already referred to (Chap. II on "Zymotic Disease"), and the same substance has been patented under the
name of Condy's Disinfectant.

The second of this class of disinfectants are the agents
which promote oxidation by a physical property, that is, by
bringing the putrid matters into contact with atmospheric
oxygen. These are fire, water, and porous solids.

The value of fire as a disinfectant is well known, and has
been recognised from remote periods. Dr. Letheby observes :—"The sacrificial altars of early nations were the
rude methods by which this agent was employed." Fire as
a destroyer of miasm has been already referred to (Chap. III
on "Malaria"), has been recommended as a purifier of the
atmosphere during epidemic cholera visitations (Chap XVII,
on "Cholera"), and also as the means of disposal of débris, in
one of the pages immediately preceding.   .

The disinfectant powers of water are great indeed, as is
evidenced by the immense masses of sewage discharged into
sewers, and which there undergo oxidation. But it has
also been demonstrated (Chapters XVI and XVII, on
"Cholera and Water"), that water may become the medium
of dissemination of disease ; and the state of the Thames
proves that the disinfecting powers which any volume of
water possesses may be too severely taxed, and that, under
certain atmospheric conditions, it may become offensive, and
insidiously destructive to health, even if not actively deleterious.

The last means of destroying offensive matter is by the
agency of a porous solid. The best examples of such agents
are common clay and charcoal. Both of these operate in

the same way, by condensing the putrid vapours within their pores and on their surface, so as to cause them to unite with oxygen, forming slow combustion, on the eremacausis before mentioned. To effect this, however, there must be free access of atmospheric air, and the substances must be dry, otherwise the power of deodorisation is reduced.

The power of common earth as a deodoriser has been already referred to in this chapter, and this knowledge recommended to be made use of for the destruction of cholera fæces. (See chapter XXI, "On Prophylactic Medicine").

Charcoal, besides expediting the oxidation of injurious matters by virtue of its porosity, is capable of absorbing immense quantities of gaseous bodies. This property of charcoal was noticed by Saussure as far back as 1814, but Dr. Stenhouse was the first to propose it as a purifier of foul gases which escape from decaying matters. Of the different kinds of charcoal, vegetable is the most effective, and this is doubtless due to the greater porosity. Liebig states that the pores in a cubic inch of beech-wood charcoal, must, at the lowest computation, be equal to the surface of 100 square feet. Hence, when this substance is exposed to an atmosphere containing the putrid products of decomposition, it absorbs and oxidises them by the species of combustion referred to " as effectually as if they were passed through a furnace." Boxes of charcoal, placed in latrines and other places where the products of decomposition abound, will absorb any gas with which the charcoal comes in contact up to ninety times its bulk, and the only precaution necessary is to keep the charcoal perfectly dry ; for when wet, it materially loses its power of deodorisation. Ewart,[1] however, remarks :—" Exposure to an Indian sun is quite sufficient to expel the moisture absorbed from the air. But if, after long use in the more humid climates of this country, or in the monsoon months, this simple measure fails, reignition for a few minutes

[1] Ewart, 'Sanitary Condition of Indian Gaols,' p. 155.

in a closed space or vessel, will always suffice to restore its original vigour."

The editor of the 'Lancet,'[1] in an article on the relation of typhoid fever and filth, informs us, that an interesting experiment has been proceeding for two years in a large and crowded district in the city of London, for the purpose of determining the value of charcoal filters on the atmosphere of sewers. Dr. Letheby has reported that, as far as his observations have gone, the results from the use of this deodoriser are most satisfactory." In examining the charcoal after two years' work, he finds it charged not only with sewer miasms in an alkaline form, but also with the products of their oxidation.

To conclude, the importance of rigid conservancy in all India cantonments cannot be too strongly insisted upon. Of this subject it may be emphatically observed—

"Though low the subject, it demands our praise."

---

[1] 'Lancet,' January 18th, 1862.

CHAPTER XVI.

ON WATER.

Composition of Water—Importance of Water in Animal Economy—Importance of examining all Water, particularly in Tropical Climates—Rain Water—River Water—Well Water—Tank Water—Marsh, or Stagnant Water—Diseases arising from Impure Water : Malarious Fever ; Goitre ; Paralysis—Action of Water on Lead : Diarrhœa ; Dysentery ; Cholera ; Guinea-Worm ; Other Entozoa ; Yellow Fever ; Stone in the Bladder ; Dyspepsia ; Gout—Tests of Water—Purification of Water : *Strychnos potatorum.*

WATER, as every one knows, or at least should be so aware of, is composed of two gases, oxygen and hydrogen, united in the proportion of eight parts by weight of oxygen to one of hydrogen. It is also the universal solvent of nature, and, forming more than half the weight of all animal and vegetable bodies, plays a most important part in the economy of life. It is employed to dissolve the materials of the food, that they may gain entrance from the gastro-intestinal passages into the blood ; in the latter fluid it holds the nutrient matters in solution, and carries them to every part of the body for the removal of its component parts; it conveys away effete and decaying material no longer required ; in short, performs a most essential part in the actions of digestion, nutrition, circulation, secretion and excretion. Hence no care can be too great to secure an unsullied supply of this indispensable necessity.

The deleterious effects of contaminated water having been so repeatedly observed, it has become important that the

11

physician and sanitarian should contemplate such contaminations as belonging to the proximate causes of disease. Moreover, it is not sufficient that the purity of the water of a district should be examined into ; but it is absolutely necessary that the non-contamination of the water used in a barrack, hospital, or residence, should be authoritatively assured.

In tropical climates, from the prolific abundance of animal and vegetable growth, it is doubly incumbent that the quality of the water receives thorough investigation. It frequently happens, especially in India, that besides containing excessive quantities of chloride of sodium, nitrate of potash, lime, salts, and other inorganic solutions, water contains remains of decomposing animal and vegetable substances, and also living organisms, which act as irritants on the intestines, enter into the tissues of the body, or otherwise influence the condition of the solids and fluids of the system.

Next to artificially distilled water, rain water, in the product of nature's distilling process, is most pure. This, however, falling on the earth, exerts its solvent powers, and as it percolates through the ground acquires those impurities, and becomes productive of those diseases now about to be briefly noticed.

Next to distilled or rain water, river water is perhaps the most free from impurity in its natural state, although, as Dr. Ranken observes, "The rivers of India, however pure on issuing from remote mountains, are more and more polluted in their course by the confluence of rivulets and smaller streams." These "nullahs" are formed by the rain, which, falling upon trees, herbs, and grass, dissolve a portion of their substance, and, in addition, serve as the common sewers of the wide regions which they traverse. They are the last receptacles of all that has ceased to live, from the bodies of men and quadrupeds and myriads of animalculæ and aquatic plants that breed and rot on their margins.

Thus it frequently happens, that river water, in addition to the ordinary twenty grains of saline matter to the gallon (consisting chiefly of carbonate of lime, common salt, and sulphate of lime), contains in a state of mechanical suspension a large quantity of other decomposing mineral and organic matter. And this is, of course, especially the case in those streams which receive the sewage and refuse of large cities.

Hence, notwithstanding pure river water is well fitted for drinking purposes, that in the neighbourhood of, or below large cities, is quite the reverse; although, from the absence of large quantities of earthy salts such water is well adapted for domestic use, as washing, soap not being precipitated by it to any great extent. It may also be used with safety for cooking purposes, the process of boiling tending to destroy all animal impurities, and to decompose and precipitate the inorganic solutions.

Well water.—As well or spring water passes through a considerable mass of soil before it collects in crevices and chasms below the surface of the earth; it is apparent that it must become the solvent of a large amount of mineral material. In percolating through the soil it dissolves all the soluble mineral matter within its reach, and hence the fact that spring water is the "hardest" of the waters in common use.

In ordinary cases it contains about thirty grains of calcareous or other salts in each gallon, as well as carbonic acid gas. If, owing to any peculiarity of the soil there is a large quantity of saline or mineral matter present, the water becomes a mineral water, and is unfit for common use.

The earthy salts in spring water decompose and precipitate soap, and thus the fluid is unfitted for washing purposes. Owing also to the larger amount of mineral matter present, its solvent power is less than that of "softer" water—as rain or river water—which contains a smaller per-centage of inorganic material. Hence it is not so well adapted for cooking purposes, although, during the boiling process, it becomes "softer," from escape of carbonic acid gas and deposition of

lime on the utensils, which, indeed, forms the "fur" house-
wives complain of on the inner surfaces of their tea-kettles.

Well water, however, being free from obvious causes of
impurity is popularly considered pure, and is universally
preferred for drinking purposes. As a general rule, unless
the amount of earthy salts be excessive, spring water may be
drank with safety. There are, however, certain complaints
which it cannot fail to aggravate, as some forms of dyspepsia,
gout, all calculous affections, and most diseases of the kidneys and bladder.

In addition to inorganic impurities, well water is frequently contaminated by inorganic matter, and then becomes exceedingly deleterious to health. If the soil through
which the feeding water percolates is highly charged with
organic constituents, the fluid is liable to be filled with
oxygen largely impregnated with such compounds. Moreover, well water is frequently contaminated in a more direct
manner by decomposing organic matter. First by the direct
deposition of dead animal and vegetable substances, as insects
and leaves, into it; and secondly, by the percolation of sewage
and other decomposing materials through chinks in masonry,
or fissures in the side of the shaft. As proof of this, the
fact of wells in the neighbourhood of church-yards containing
nitrates and other results of the decomposition of animal
remains may be referred to. In like manner, neglect of
conservancy will lead to contamination of water in wells.

Tank water becomes rapidly contaminated by decomposing
organic matter, especially when, as is too often the case in
India, filth and vegetable growth is allowed to accumulate on
the margins, or in the hollow of the receptacle. The solution
of such organic matter is greatly facilitated by the comparatively small proportion of earthy and alkaline salts origi-
nally present in such water.

In marsh or stagnant water the impurities are chiefly dissolved organic matter, with the different forms of algæ and

infusoria. It is in such fluid that the germs of guinea-worm, hydatids, tape-worm, and other entozoa are most probably generated, and it is from such water that carburetted and sulphuretted hydrogen and carbonic acid gases are evolved. There is also good reason for stating that stagnant water absorbs malaria.

The chief diseases which are known to arise from the use of impure water are—malarious fevers, goitre, paralysis, diarrhœa, dysentery, cholera, guinea-worm, other entozoa, yellow fever, stone in the bladder, dyspepsia, gout.

These diseases are of a widely different character, and each springs from agents of a peculiar nature, some organic, others inorganic, but all conveyed into the system by the medium of water; and, according to their composition, these agents are either mechanically suspended or chemically dissolved in the fluid.

That malarious fevers are produced, amongst other causes, by the use of contaminated water, there is abundant proof. M. Bondin records that, in 1834, eight hundred soldiers arrived in Marseilles, from Bona, in three transports. On board one ship ninety-eight men were taken into hospital, thirteen of whom died from remittent, with typhoid symptoms. On inquiry being made, it was ascertained that on board the affected ship, the water supplied for the soldiers, owing to haste on embarkation, had been taken from a marshy place near Bona, whilst the crew, who were supplied with wholesome water, did not furnish one case of fever.

Mr. Cornish,[1] in his medical history of the Shervaroy Hills, states that the hill tribes have a great dislike to the low country. The water of the plains, they believe, causes fever.

Mr. Bettington,[2] of the Bombay Civil Service, has latterly

---

[1] 'Madras Quarterly Med. Journal,' October, 1861.
[2] Bettington, "On the Water of Nullahs in Jungle Districts;" 'Bombay Government Records,' No. 20.

brought forward good reasons for concluding that the use of stagnant water containing decomposing vegetable matter, as is the case with many of the wells and tanks of India, is an exciting cause of malarious fever.

Dr. Mackinnon also states that fever is occasioned by drinking stagnant water.

Within my own knowledge, an officer out shooting large game was cautioned by his "shikarree" against drinking water from a stagnant pond. The officer, disregarding this prohibition, drank copiously, and, as his servant prophesied, within three months suffered a severe attack of jungle fever, for which he was ultimately obliged to take a furlough to Europe.

Instances of a similar nature are frequently observed by Indian officers, many of which I have heard related.

Dr. Pidduck,[1] bearing in mind the theory of Linnæus referred to in the chapter on "Malaria," that intermittent arises from drinking water impregnated with argillaceous soil, dwells upon the practical application of the idea, and states that he has succeeded in curing intermittents and remittents, simply by interdicting the use of any but distilled water.

Marsh water is also stated to induce cachexia in sheep; and in Albania the shepherds are particular not to allow their charges to drink otherwise than from well-known healthy sources.[2]

That impure water induces the formation of wens and goitre is unquestionable. The hill tracts of Derbyshire, the Swiss mountains, some of the Indian ranges, the limestone hills in south Staffordshire, in all of which the water is more or less impregnated with gypsum and magnesian salts, may be quoted as localities where the bulky condition of the

---

[1] 'Lancet,' May, 1861.

[2] Dr. Marsden, "On Ague on Board Ship." 'Edin. Med. Journ.,' February, 1862.

thyroid gland, or goitre, or even cretinism is frequently observed.

The inmates of the Durham jail furnished a forcible example of the foregoing. In 1843 a well was sunk for the use of the prisoners, and during the time it was used—ten years—glandular enlargement about the neck was common among the patients. In 1853 the inmates were supplied with water from the Weir, and the disease quickly declined.

Dr. McLelland's work on the 'Medical Topography of Bengal and the North-west Provinces,' proves that there is an intimate relation between the occurrence of limestone in the sources from which the inhabitants of both the plains and hill tracts of India derive their drinking water and the prevalence of goitre and cretinism. On those calcareous soils, called by the natives "bhat" lands, yielding twenty-five per cent. of carbonate of lime, dogs and cats are also affected with goitre.

It is very probable that, besides gypsum and magnesia, a superabundance of any solid matter in drinking water may lead to the development of goitre, and that thus the disease, although most frequently observed in connection with the above-named formations, is not always the result of a particular salt. As a rule, however, this disease is most usually found on a calcareous soil; and it appears that this connection has been latterly further demonstrated by Kölliker, who found that in cretins the base of the skull becomes, at an early period, unusually ossified and thickened, thus leading to disease of the brain, by narrowing the foramina of the nutritious arteries. Kölliker supposes the excess of lime, introduced into the system through the medium of the water, leads to this ossification.

Paralysis.—Of the power of water in conveying the poisonous salts of lead into the human system, and thus giving rise to lead-colic, or paralysis, there are, unfortunately, too many examples. The danger of such pipes was indeed known

to Galen and Vitruvius, and, in later days, Dr. Christison
relates the history of an epidemic, which occurred some forty
years since in the town of Tunbridge, from the use of leaden
pipes.   But it will be needless to multiply either individual
or collective examples of disease arising from drinking water
impregnated with lead.   During the three years I was resident surgeon at the Queen's Hospital, I suppose two or three
such cases were always under treatment, and every medical
man who has seen much public or private practice must have
met with instances of a similar nature.   To such, the peculiar drop of the wrists, the blue line round the teeth, and the
expression of the patient, are quite sufficient to declare the
character of the disease.

According to Dr. Hassall, lead in contact with water is
oxidized in three ways : through the oxygen of the air;
through the decomposition of certain substances present in
the water, as the nitrates ; and by the decomposition of
water.   Thus hydrated oxide of lead is formed, a substance
in itself not soluble in water, and, therefore, innocuous ; but
meeting with carbonic acid, more or less present in all
waters, it forms a soluble and highly poisonous compound.

Diarrhœa.—Organic or inorganic matter suspended or dissolved in water, will alike induce diarrhœa.   Most people
who have drank the water of Aden, Nusseerabad, Beawr,
Ajmere, Mullye, and similar soils, will admit its power in
inducing the affection.   Mr. Cox,[1] in his paper entitled
'Epidemics and their Every-day Causes,' gives some striking
instances of diarrhœa from organically impure water.

Dysentery.—The same impure water which will induce
diarrhœa is also capable of exciting even dysentery.   Numerous instances of dysentery arising from contaminated water
might be quoted.   The disease prevailed in the West India
islands[2] so long as the troops were supplied with water ren-

---

[1] 'Sanitary Review,' October, 1858.
[2] Op. cit.

dered impure in its course, and ceased altogether so soon as the same water was kept free from impurities by being conveyed through a well-constructed aqueduct; and it is notorious that the disease is most rife during a period of drought, when, from scarcity of water, the inhabitants of any district are compelled to use that from ponds and other stagnant collections.

That the germs of cholera may be conveyed through the medium of water there exists, unfortunately, little doubt. Dr. Snow's conclusive argument is well known. Sixteen sub-districts of the metropolis, from Southwark to Camberwell, received their water supply from two sources: from the Southwark and Vauxhall Company, which collected its supply at Battersea Bridge, and which fluid contained an immense quantity of the sewage of London; and from the Lambeth Company which obtained its supply at Thames Ditton, a pure source. Thus with regard to drinking water, the thousands inhabiting the said districts were divided, during the cholera epidemic, into two classes, one being supplied with water containing the sewage of London, the other having water free from such impurity. Cholera was fourteen times as fatal among the persons having impure water as among other groups otherwise subject to the same influences. It is not, of course, asserted that impure water is the cause of cholera, but that the germs of the disease—the decomposing cholera fæcal *evacuations*—may be conveyed into the system by the medium of water, and thus directly excite choleraic symptoms. Moreover, as Dr. Sutherland states—"The use of water containing *any* organic matter in a state of decomposition is one *predisposing* cause of cholera, and also very much aggravates the result." It is stated in the 'Allahabad Gazette' that at Mean Meer, where last year 500 soldiers fell victims to cholera, the well water has been found to contain sulphuretted hydrogen. (For further remarks on cholera, see Chap. XVII.)

Guinea-worm.—Mr. Carter,[1] of Bombay, discovered that minute worms, having great resemblance to the young of the Guinea worm, existed in abundance in some of the ponds near Bombay, and the same had also been previously noticed at other places by different observers, as Messrs. Duncan and Forbes. Although some investigators, as Mr. Greenhow,[2] of the Bengal service, have not been able to find these worms in other localities where dracunculus prevailed, still the weight of evidence appears in favour of the supposition that the worms in question are the young of the guinea-worm, perhaps slightly *altered by different circumstances* of existence. There is, moreover, good reason for the belief that the guinea-worm may effect an entrance from water into the human system in two ways, viz., by the mouth, and *viâ* the sudoriferous channels. This question, however, is foreign to the purport of the present work, but is fully treated of in my ' Manual of the Diseases of India. It may, however, be remarked that as by "embolism," Professors Virchow and Cohn account for the conveyance of substances detached not only from the valves of the heart to the cerebral arteries, but also from a far off nidus amongst diseased tissues, whence such substances travel along the veins *to the heart*, so in like manner may the germ of the guinea-worm be conveyed from its entrance by the lactiferous ducts of the digestive organs, to all parts of the body. And that the germ may escape the action of the latter is evident from the following :

Other Entozoa.—Elliotson relates the case of a whole family becoming affected with ascarides from using the water of a well in which minute worms were subsequently found, and losing the disease when they discontinued drinking the water. The same author relates the case of a child who discharged numbers of maggots after eating "high pheasant." The fact of the gadfly arriving at maturity in the stomach

[1] Carter, "Notes on Dracunculus;" 'Trans. Bomb. Med. Phy. Soc.,' 1853.
[2] Greenhow, ' Ind. An. Med. Sci.,' vol. vi.

of horses is also a well-authenticated fact. Infusorial, ciliated organisms, existing in water, have latterly been asserted to be the embryos of the *Bothriocephalus lata,* or broad tape-worm.[1]

The French army on entering Syria suffered from drinking water containing the Sanguisaga Ægyptica, a very thin and small leech, which existed in the muddy pools. Caterpillars, larvæ of flies, and other living bodies are occasionally introduced through the medium of water.

Yellow Fever.—On the subject of yellow fever in relation to water supply, but little can be said of a positive nature. It is, however, certain that wherever yellow fever has appeared there has the water been generally in a filthy and faulty state ; and much bearing on the subject, and leading to the proof of the above assertion, may be found in ' Papers respecting the Origin of the Yellow Fever in Bermuda, in 1853.' It is there stated that the tank water which supplied the barracks in Fort Cunningham became so offensive that it could not be used, and the great mortality which took place in that quarter of the barracks was attributed to the use of this water and the effluvia generated.

In the town of Bermuda itself there were no proper arrangements for the supply of water at all. In the majority of cases, the water is stored in tanks, and impregnated with sulphuretted hydrogen. When to these facts are added the statements that in Bermuda drainage and sewage are neglected altogether ; that the contents of privies and cesspools are allowed to accumulate in natural pits and hollows, and on the ground surface, and that the tanks are filled with water percolating through this impure soil, there is not wanted much further evidence to prove that the effects of the vile water, which must at the time of the pest have been hourly consumed, had no small share in *intensifying,* if not producing the disease.

---

[1] ' St. Petersburgh Medical Gazette,' Jan., 1861.

Stone in the bladder.—Wherever the water is overcharged with saline matter, particularly lime salts, calculus vesicæ is a prevalent disease ; as examples, may be instanced some parts of Derbyshire, portions of South Staffordshire, the Province of Kutch, in Western India, Hyderabad in Scinde, and many other places.

That dsypepsia and gout will be aggravated by the use of impure water, particularly that containing certain inorganic material, is sufficiently evident to all who understand the nature of the diseases named.

Pure water should remain limpid on the addition of lime-water, chloride of barium, nitric acid, oxalate of ammonia, or hydro-sulphuric acid, thus showing the absence of carbonic acid, sulphates, chlorides, and lime salts.

If peas or beans be boiled in water containing carbonate of lime, the carbonic acid is expelled, and the lime precipitated, incrusting the cooking vegetables, which, therefore, remain hard. Resinous substance also will not mix with water containing lime, and soap is decomposed in the same fluid, the alkali of the soap uniting with the acid of the earthy salt, while the oil and earths combine to form insoluble mixtures, which swim on the surface. Hence these latter may be called " ready methods" of judging of the "hardness" or " softness" of water.

The soap test may be otherwise applied in a sufficiently easy manner. If a drop or two of solution of soap, in ether, be added to a test-tube half filled with water, a greater or less opacity will be produced, according as the fluid contains more or less lime.

The ammonia test may also be very readily applied. Drop into the specimen one drop of solution of ammonia, and then two drops of a solution of oxalate of ammonia. The lime present will be thrown down in the form of a white oxalate. The formations from these two tests should be compared with the results from the addition of the same to distilled water, by which an approximation may be made by the eye as to the quantity of earthy salts contained in the fluid.

Purification of water.—Boiling water destroys confervæ, infusoria, the germs of guinea-worm, and other entozoa, and also will probably render organic matters, as the cholera germ, innocuous. The same process precipitates lime, and never ought to be neglected where contaminated water is obliged to be used.

If the water be afterwards passed through a common sand and charcoal filter, which every bhistee knows how to construct, it will be cleansed from all impurities, and be pleasant to the taste. Such a plan should be adopted in most barracks, hospitals, and residences in a tropical country. It would be attended with little expense, care, cleanliness, and a few "gharras" (earthen pots), filled with sand and charcoal, being all that is required.

The fruit of the Strychnos potatorum is made use of by the natives in some parts of India, to cleanse their water. Its Hindoo name is "Nirmulee," and it grows chiefly in the forests of Southern India. Unlike other varieties of strychnos, it has no poisonous properties, and the cut fruit being rubbed on the side of a vessel is sufficient to clear the most muddy water in a few minutes. Dr. O'Shaughnessy[1] thought this property due to its astringency, but Dr. Pereira[2] has latterly shown that it depends on the presence of albumen and casein.

The common bazaar alum called "phitkare," is also very useful as a purifier of dirtied water.

If water from muddy ponds or tanks must be used, the following is a good and ready method of obtaining it partially purified :—Place a barrel in the water, the sides and bottom of which have been previously perforated ; in this barrel let a sheet containing fine sand and charcoal be suspended. The

[1] 'Bengal Dispens.,' p. 443.
[2] 'Phar. Journal,' 1850, p. 478.

water passes through the barrel and sheet, and may be obtained tolerably clear from the surface of the filtering mixture.

A small filtering stone, to which a flexible tube is attached, is a very convenient article for private use. It is manufactured as a pocket case, and the stone being placed in water, the liquid may be sucked through the tube, quite free from all organic impurities. On a campaign this is an exceedingly useful article to carry about the person.

It is stated that the country onion is a safeguard against thirst.

CHAPTER XVII.

ON CHOLERA.

Sporadic and Malignant Cholera: Origin and Dissemination; Meteorological Causes; Local Causes—Cholera Infests the Ground—Cholera Fœces will induce Cholera—Water a Medium of Cholera—Cholera in the Black Sea Fleet—Want of Sanitation causes Cholera—Dangers of Sewers beneath Dwellings: Mr. Simon's Opinion—Prevention of Cholera—Sanitation insisted on—Movements of Troops insisted on—Cholera in India follows the great lines of Human Intercourse: Deduction—Good Results and Dangers from Railroads—Hindoo Pilgrimages a Cause of Cholera—Measures to be adopted when Troops cannot be marched away from Cholera—Officers not so liable to Cholera as the Men.

SPORADIC cholera appears to have occurred in all ages. Hippocrates[1] enters into details respecting its history and symptoms. Aretœus[2] gives an admirable account of the disease. Celsus[3] also delineates its most striking features.

The malignant or Asiatic form appears to have first shown itself in India, at Madras, in 1774.[4] In 1780 it destroyed 20,000 persons assembled at the Hurdwan festival. In 1810 it occurred at Jessore,[5] and ultimately spread over the whole country. Since that period it has always prevailed with more or less intensity, and under the most varied conditions of climate, at the level of the sea, and in the Himalayan range, desolating small villages and populous towns, sparing neither age nor sex.

[1] 'De Morbus Popularibus,' lib. v.
[2] 'Aretœus,' lib. ii, chap. v.
[3] 'Celsus,' lib. iv.
[4] Curtis, 'Diseases of India.'
[5] Roupell, 'On Cholera.'

And yet, notwithstanding the varied circumstances under which the disease has existed, the laws under which it originates and pursues its course, and hence its preventive treatment, are much more satisfactorily demonstrated than the method of cure when the disease is present.

Isolated outbreaks of cholera cannot leave any doubt that the disease may arise without communication from any infected source. Such occurrences are only explicable by the fact, that the poison of cholera is capable of *production* under favourable circumstances. This, of course, does *not negative the transference* of the poison, which may thus, forming independently, be the means, on being removed to a fresh locality possessed of the requisite local conditions, of again exciting a similar form of decomposition; and the meteorological conditions which have been so frequently found to precede or accompany epidemic visitation of cholera are undoubtedly favorable to the particular form of decomposition which gives vigour to cholera poison.

Hence the circumstances under which epidemic cholera has occurred are referable to two heads, meteorological and localizing causes. Allowing for differences in climate, certain meteorological phenomena have most generally accompanied outbreaks of cholera; the chief of which conditions have been found to be a somewhat variable, but elevated temperature, a still and peculiarly oppressive state of the atmosphere, more oppressive than simple elevation of the thermometer can account for, conjoined with a certain degree of moisture, and frequent absence of ozone.

Seasons, however, presenting all the characteristics of what may be termed the choleraic pestilential constitution of the atmosphere have existed very frequently where there has been no accompanying or following pestilence.

Another co-efficient at least is therefore required, in order to give character and energy to the seasonal conditions which favour the development of cholera, and this is to be found in local causes. That such cause is strictly local, is evidenced by the fact that an analysis of the history of cholera epidemic shows them to be most frequently made up of a *succession of partial local outbreaks*, and this not only as regards different regions, but also in the same place. Thus, although whole districts partake of the same meteorological influences, some usually escape an epidemic visitation, at the very period when others in the immediate vicinity are suffering severely from the disease. Even in the same town or camp, when the inhabitants of some streets or lines of tents are being decimated, those dwelling in others not far distant altogether escape; or, as frequently happens, the inmates of certain houses suffer severely, while their neighbours are entirely spared.

Nay, more, in my own experience, it has occurred that choleraic diarrhœa has been confined to those sleeping on one side of a troop-ship, where temporary side houses had been rigged outside the vessel for the accommodation of the soldiers; the same quarter being exposed during a voyage of ten or twelve days to the prevailing monsoon breeze. In this case the diarrhœa ceased on the removal of the side houses of the ship. Had we been under an epidemic condition of atmosphere, or had foul weather obliged us to close *most* of the ports, there is little doubt that cholera would have resulted.

In former years we have the authority of Trotter to the effect that the stench from the head entering the sick bays of our ships was an evident cause of disease, especially diarrhœa and typhus.

Thus, as Dr. Farr observes, the intensity of cholera depends " both on local and meteorological circumstances."

Among the local causes which have been demonstrated as developing an outbreak of cholera we have lowness of level, bad arrangement of streets, houses or tents; defective ventilation in the same; want of cleanliness, over-crowding, dampness, impure water, effluvia from the decomposition of the various organic débris allowed to collect in poor and neglected localities; the emanations from human or animal excrement, whether accumulated in cesspools, or allowed to rot in foul drains; the malaria from fetid ditches; the miasm from city grave-yards; the stench from knacker's yards, and from some offensive mercantile establishments, as catgut spinning and other injurious trades.

The general rule that the mortality of cholera is inversely as the elevation of the people assailed above the sea level, is demonstrated by Sir Ranald Martin, and at a still later period by Pettenkofer, who states that "cholera prevails more intensely in the low districts, because, all the organic impurities of the higher grounds gravitating thither, these undergo chemical action." The latter author[1] also entertains the belief that cholera never prevails epidemically on rock, and accounts for the supposed fact on the principle that excrement cannot penetrate into the soil, and that rock neither absorbs nor gives off moisture.

Those authorities who have not ventured to ascribe the cause of cholera to any particular local error, do not hesitate to class it as one of those diseases which affect the ground. Thus, Mr. Simon[2] states "cholera is one of those diseases which infect the ground." Budd[3] states :—" Like malignant cholera, dysentery, yellow fever, and some others that might be named, this (yellow fever), is one of the group of diseases that infect the ground;" and Dr. Greenhow[4] says, not only did cholera manifest a liking for particular localities, but in those localities it affected particular houses; and Indian reports afford ample evidence that cholera has been partial to localities where the atmosphere was vitiated by the products of

---

[1] 'Investigations on the Propagation of Cholera.' Munich, 1855.
[2] " Disregard of the Laws of Health ;" 'The Times,' June 17, 1861.
[3] ' Lancet,' July 23rd, 1859.
[4] " Report on Cholera in Tynemouth ;" 'Sanitary Review,' June, 1855.

fæcal decomposition deposited on the ground; as at Jessore in 1817, at Meerut in 1819, at Vipery, and many other places. That the excretæ of cholera will produce cholera, receives very convincing proof from the experiments and researches of sundry investigators. Becker says, in 1853, four young physicians tasted the cholera evacuations, and each suffered from choleraic manifestations. M. Pinel inoculated himself, and suffered, I believe, from diarrhœa. M. Foy tasted the fæces, and had the same disease. An assistant in the Clinical Hospital of Berlin tasted the fæces, and had cholera. Dr. Thiersch, of Munich, shows that cholera excrement, although not at first capable of producing cholera, did bring on symptoms of the disease, *after being kept six or eight days,* and then administered to animals. Dr. Lindsay's experiments are also affirmative.

On the other hand, instances have occurred where the disease was not induced by swallowing cholera fæces, and Dr. Snow refers to the case of a dispenser at Newcastle, who drank by mistake some rice-water evacuation, but was not attacked.

It seems probable that, as Dr. Thiersch states, a certain amount of *decomposition* of cholera fæcal matter must occur before the material will induce the disease. Much, however, would certainly depend on the state of health, both morally and physically, or the presence or otherwise of the *mens sana in corpore sano* of the individual experimenting.

That impure water has a powerful influence over the intensity of cholera outbreaks is unquestionable, and has already been noticed in a former chapter (Chapter XVI, "Water"). Dr. Sutherland, in 'A Report to the Board of Health,' states : —" A number of the most severe and fatal outbreaks were referable to no other cause except the state of the water supply," and this especially where the water was obtained from wells into which the contents of sewers, privies, or the drainage of grave-yards had escaped. Dr. Acland[1] also re-

[1] 'Memoir of Cholera in Oxford.'

cords two valuable examples; one in the parish of St. Clement, the other in the county gaol. The researches of Dr. Snow on this point have been already given, and appear almost conclusive that water will convey one at least of the *materies morbi* of cholera.

From the report on cholera in the Black Sea fleet, it appears that on the water, as everywhere else, the most crowded, worst ventilated, and, in all respects, least sanitary places generally suffer most. Dr. Rees, surgeon of the " Britannia," attributes the outbreak on board that vessel in a great measure to defective ventilation, and adds that, when a return to port was decided upon, the continued violence of the scourge, the crowded state of the middle-deck, the discharge from the bowels and stomachs of the sick, and the continued want of adequate ventilation, had continued to render the ship a laboratory of pest-poison.   In the " Albion," 419 cases of diarrhœa and cholera occurred, among a crew of 800, of which 69 proved fatal.   The surgeon accounts for this mortality because the evacuations from those previously affected had accumulated and acquired a much more deadly influence from imperfect ventilation, in consequence of the state of the weather.   Supposing that the earlier cases had received the disease on shore, and that their discharges accumulated between decks in the manner named by Mr. Rees, and the surgeon of the " Albion," there is no longer any difficulty in understanding how, in the hot, close, confined, still atmosphere of a ship's lower-deck, miasm would arise and become concentrated, just as was formerly the case in gaols and prisons, and so induce the subsequent disastrous outbreak.

Having thus found that there are numerous localising causes of cholera, the question presents—are any or all of these necessary to the existence of the disease?   All of them certainly are not so necessary, and frequently it may appear that cholera has arisen without *any* such concomitants.   I believe, however, that the scientific physician who has studied the laws and habitats of cholera, will but rarely be at fault, in fixing upon the local cause, although such local cause may be hidden from, and, in fact, not admitted by the public; may require meteorological conditions for its development; or may have been presented miles from the place where it developes itself in cholera.

Thus we find individuals dying of cholera, who, living without sanitary defects in their immediate locality, imbibe the disease in their walks or drives abroad. They pass through a locality ninety-nine times, and the hundredth their system happens to be in a particular condition predisposing them to disease; they take cholera or other malady, and die victims to a want of general sanitation.

The result of all observation has been, that an atmosphere impregnated with the products of fermenting excretæ, is at once one of the most obvious, and most constant concomitants of cholera. Such exhalations have often been found in houses where the existence of any palpable cause of insalubrity would scarcely be suspected, and thus the fact of the pestilence sometimes passing over the houses of the poor, and appearing in the mansions of the rich, may be explained. The crevice in the drain, the crack in the sewer, passing beneath the house, insensibly contaminates and poisons the air.

Mr. Simon in his 'Fifth Annual Report to the Commissioners of Sewers,' writes :—" Through the unpolluted atmosphere of cleanly districts it migrates slowly without a blow; that which it can kindle into poison lies not there. To the foul damp breath of low-lying cities, it comes like a spark to powder. Here is contained that which it can swiftly make destructive, soaked into the soil—stagnant in water, griming the pavement, tainting the air, the slow rottenness of unremoved excrement, to which the first contact of this foreign ferment brings the occasion of changing into new and more deadly combinations."

It is indeed true that the chemist has been unable to detect the peculiar poison of cholera, or the peculiar decomposition which is its exciting cause; but he is equally at fault, as has already been shown (Chap. III, on "Malaria"), regarding the detection of the paludal poison, or in discovering the nature of the peculiar conditions of which typhoid poison is the product.

As, however, malarious disease, or degeneration will certainly be induced in the system of the human being subjected to such influence, so it is equally certain that cholera and kindred affections will arise under certain other conditions of life. It has long been known that, under such conditions—the inspiration and absorption of an atmosphere impregnated with animal decomposition—the alimentary canal becomes disturbed, and fluxes occur. Dr. Cullen, years since, remarked that the effluvia from very putrid animal substances readily produce diarrhœa—an observation confirmed by daily dissecting-room experience, and by the prevalence of diarrhœa and cholera in those localities where decomposition abounds.

If there be one particle of truth in the foregoing remarks, the prevention of cholera in India—as everywhere else—depends upon the adoption of a thoroughly efficient sanitary system, and on the immediate removal of all decomposing matter, which can possibly result in the formation of the cholera poison, *or add to its virulence* on its approach from another quarter. *It is not attention to any one particular sanitary matter which will prevent the epidemic prevalence of cholera;* it is a thorough regard to all the dictates of sanitary science, which will render the epidemic intensity of the disease impossible.

Hence it is not sufficient that an Indian cantonment be kept merely *apparently* clean; the prevention of cholera demands more than this. Scavengers should be superintended by a competent officer, refuse disposed of as recommended in Chapter XV of this book, drainage of the whole locality thoroughly effected, holes and hollows filled up, and no collection of rubbish allowed to exist, bazaars placed under strict sanitary supervision, latrines erected and their use enforced, the state of the barracks, canteens, soldiers' dormitories, wash-houses, urinals, guard-room, diet, means of cooking, and places of recreation narrowly inspected, duty lessened short of fatigue, and the mind occupied and kept in a healthy state, free from anxiety regarding the approach of the disease—in short, the prevention of cholera demands the strict application of all the recommendations made under the different heads which compose this volume.

When cholera unhappily prevails in a district, carriage, tents, and commissariat supplies should be ordered, so that on the first appearance of cholera at a cantonment, the whole body may change ground, leaving their barracks for positions assigned to them in different directions. It may not be necessary to move out more than a few miles; but this should be *done at first*, without delaying, and thus allowing the disease to depress the spirits of the men, which occurs at times to so great an extent, that, as was the case in September, 1861, at Lahore, suicide becomes almost an epidemic.

Listlessly to await the approach of cholera, to allow soldiers to brood over the coming danger, without healthy excitement or distractions to lighten their minds and divert their attention, *is simply to prepare them for death.* If, however, the initiative be taken—if the soldiers, instead of remaining idle, and dejectedly awaiting the steadily approaching pestilence with vague alarm, are busied with preparations for moving— the prospect of change and excitement will raise the spirits of the men, and prepare them to resist disease. This course has been pursued so many times with success, that arguments in its favour are not required. As when yellow fever attacks a ship's crew, standing orders should direct the vessel's course to northern latitudes, so, on cholera approaching an Indian station, the rule should be, *evacuation of such cantonment by European troops.*

It is a well-known fact that cholera in India has appeared to follow the lines of human intercourse, the great rivers and roads stretching through the vast extent of territory.

It is, therefore, incumbent, on marching troops, to avoid such routes, otherwise the contaminated ground will surely give rise to the disease. Dr. Chevers[1] states:—"As a general rule, subject, unhappily, but to few exceptions, outbreaks of cholera may be anticipated whenever it becomes necessary to move troops in Lower Bengal, whether by boat, by carriage, or by marching." And so, indeed, it will ever be, as long as the great lines of human intercourse are contaminated by the crowds of natives who travel thereon, without the slightest attention to sanitation, or even cleanliness; as long as the large rivers are made the receptacle of the refuse from hundreds of cities; so long will our troops, travelling by such routes, become the prey of epidemic disease. It is not possible to reform the habits of the natives, it is not possible even to reform the sanitary condition of their large cities, and, therefore, the only course left is to avoid the more frequented routes, whenever this may be at all practicable, and the railways are fast enabling us to do this.

While, however, the latter may empower us, in many instances, to move our troops without attacks of cholera *en route*, they would appear also to disseminate disease, as a communication to the 'Times of India,' November 23, 1861, now quoted, demonstrates. That the yearly concourse at the various shrines of India tends to originate and spread disease, particularly cholera, there can be no doubt; and nothing that I can write will add force to the following lines. Let it be recollected that Punderpoor is only *one* of many similar resorts of religious enthusiasts!

"The general public has very little conception of the con-

---

[1] 'Indian Annals of Med. Sci.,' vol. x, p. 702.

cern they have in the scenes of Punderpoor. Hundreds of
miles it may be from them; it is nevertheless that festering
social spot which yearly gathers, and then bursts, on all the
surrounding country. I allude to the plague of cholera, which
is a scourge rarely absent from Punderpoor at this season of
pilgrimage. On the present occasion, the number seized was
very great, and well on to 300 are believed to have died on
the spot in seven days' time. Fatigued with long marches
of 50, 100, 200, and even 500, in certain cases 1000 miles,
hurt by indifferent food and the use of bad water on the way,
the poor pilgrims reach the spot very often with fainting
hearts, weak and aching frames, swollen feet, and emptied
pockets. They encamp in the sandy bed of the river, for the
most part, out of which vapours are doubtless ever reeking
into their little tents. When all are gathered, there is a
whole day's fast observed, as if to diminish the chances of
life to the utmost. Every cause, moreover, contributes gradually to fill the air with poison and pollution. The influx
of 50,000 to 100,000 people, their close aggregation, and
their habits and necessities, soon contaminate the whole
atmosphere; they are quickly seized and carried off by
cholera, and either burned or buried close to the scene of
their death, and in the view of the thousands who are running the same terrible risk. On the present occasion, from
which I have just returned, the sand of the river-bed was a
kind of cemetery; it was occupied partly by the living and
partly by the dead.

" Had bullets been flying about, the results would not have
been more fatal. A severe engagement on these sands between contending armies would not have left more, perhaps,
for burial on the spot than the pestilence was doing a few
days ago. The wretched sufferers seemed to have the vital
powers mysteriously checked; their pulse was generally gone,
and their eyes were horribly sunk in their sockets. Families
in which a breach was made by death sometimes fled-panic-

stricken ; and yet the orthodoxy of Punderpoor teaches that
death there is the sure road to heaven. A degrading famili-
arity with death, and a rough treatment of the dead bodies,
struck me as painful features in the sad scene. Horrid sights
witnessed still flit before my imagination. Some of the dead
appeared to be buried in a sitting posture, and destitute of
covering; none seemed to be allowed a grave of the full
length of the body. The funeral piles of those castes which
burn were ever blazing, and the sides of the river were full
of extinct but recent piles. The odour of the burning dead
was another occasional grievous offence in the atmosphere.
A more sad and mournful scene could hardly be imagined.

" To the horrors of the place must be added the horrors
propagated in all directions from it. The routes of the pilgrims homeward are ravaged by the disease, and the villagers
along them fall victims as well as the pilgrims themselves.
In all the villages along my homeward road at which I inquired, I found cholera to have been and to be at work. On
reaching the railway, I found they had carried it to Sholapoor, and the very train which brought me to Poona contained
cases which ripened on the way so as to end in death at the
Poona station. Poona and Sholapoor are now, on the return
of the pilgrims, much more full of fatal sickness than they
were. The same influence is at work all over the country,
and it is here the public has so much interest, in respect of
both health and life, in the Punderpoor jattra. They should
look to their interests; and the fatal consequences to the
whole country side would form a most just ground for the
active interference of Government, to prevent, as far as possible, not only the self-destruction of hundreds, but the
devastating ruin which they spread far and near, in every
direction. The seeds of disease are sown broadcast over the
land, and many months sometimes elapse ere it is eradicated
or spends itself out. Our cities and cantonments are in yearly
and imminent danger from the clouds of disease which are

wont to gather at Punderpoor, and then pour themselves over the whole of the Bombay side of India.

"That men should be senseless enough to leave their homes, neglect their avocations, ruin their health, waste oceans of money (not less, probably, than a million of rupees yearly in the total), in order to worship a senseless image of black stone, known as Vithoba, and wash away their sins in the Bheema (called at Punderpoor the Chandrabhâjà), may be bad enough. That they should worship a being of limited knowledge, and an adulterer—which the stories in vogue regarding Vithoba make him out to be—may also be a painful proof of man's utter stupidity and depravity. The follies and evils of Punderpoor, moreover, may be enough to disgust right-thinking men with the heartless and unworthy course pursued by the youth of India generally, who have received the benefit of a measure of education, who *practically* yield the whole weight of their influence to the support of this shocking system, and who, trembling at the consequences, refuse to withdraw from connection with these systems of wicked deception, striving to varnish them over with a worthless philosophy. All this and much more may be most proper both to think and say; but the great matter, Mr. Editor, which I should wish, for the interests of the public, to be considered is, whether annual inundation of the country with disease and death from the bed of the Chandrabhâjà, is compatible with the general safety and welfare? Let death at our doors answer the question. So long as this vile and degrading pilgrimage continues, we shall have the certainty of cholera appearing from time to time. There is more need than ever now to look to the matter, as the railway communication tends to increase the number of pilgrims. It has done so already in the yearly aggregate; but it may possibly yet increase the crowding on the great occasions. Vithoba's fame is rising; and I fear the highest inventions of our age may, in present circumstances, do no more than facilitate Hindu superstition."

The 'Poona Observer' also, of December 3rd, 1861, states:—

" We hear by letter from Punderpore that there were more than 40,000 pilgrims collected together this yea rat the fair.   Deaths from cholera are reported at forty daily, and the managing committee of the God Vithoba are in trouble, and for trial before the Mamlutdar of Punderpore."

If the observations previously referred to, of Professors Thiersch[1] and Pettenkofer[2] are recollected, and which appear to prove that cholera evacuations undergoing decomposition will excite the disease, the diffusion of the contagion, by returning pilgrims, becomes sufficiently evident.   As Professor Pettenkofer remarks, an "infective influence is exerted by the fæces of such persons in the soil of ill-conditioned places to which they go."   An infection of this kind would probably extend itself to the polluted well waters of such soils, and would render them capable of exciting cholera. Dr. Gibb informs us that an epidemic of smallpox was the result of the opening of a cemetery at Quebec, where, years since, a large number of patients, the victims of variola, had been buried.   In similar manner there is every reason to believe that cholera germ may remain quiescent for an indefinite period, until called into vitality by meteorological and atmospheric conditions.   It should be recollected that the fæces of choleraic diarrhœa may be sufficient to excite the disease, and that decomposition of fæces may commence even before they leave the body.   Probably, even ordinary fæces during an "epidemic period" may, as suggested by Mr. Simon,[3] acquire a specific infectious property.

---

[1] Thiersch, 'Infectionsversuche an Thieren mit dem Inhal. des Cholera-dermes,' 1856.

[2] Pettenkofer, 'Über die Verbreitungsart der Cholera,' 1854.

[3] Simon, 'Sanitary State of the People of England,' 1858.

On those melancholy occasions when, from military neces-sities or other paramount causes, troops cannot, as was the case with the 52nd regiment at Lucknow, be marched away from the disease, certain measures should be adopted, which experience shows have, in former instances, tended to check the epidemic.

Fires should be lighted in the neighbourhood of barracks and cantonments, their action being, as referred to in the chapter on " Malaria," to purify the air, and destroy morbid material.

The diet should be more especially attended to, and matters likely to induce abdominal irritation prohibited. An extra dram of spirits may also be allowed at night with advantage.

Employment and amusement *short of fatigue* is impera-tively demanded; work should be lessened, and amusing employment increased during visitations of cholera.

The propriety of confining the men to barracks during cholera epidemics is very questionable. The fact of ordering them to remain in barracks, or within certain limits, tends to depress them, and thus renders them more liable to be attacked by the disease. It will be sufficient that they should be prohibited visiting those spots were cholera is known to exist, in bazaars or otherwise; and that they should be cautioned against wandering into those localities which the sanitary officer points out as unwholesome.

During cholera visitations prophylactic medicine should be had recourse to, as recommended in Chapter XXI, on that subject. The method of neutralising and disposing of cholera excreta there suggested should be particularly attended to.

It is a fact well authenticated, that officers in India are less liable to cholera than the men under their command. This arises from the more satisfactory sanitary conditions under which the former live; from their being surrounded by greater comforts; committing, as a body, less excess; being less debili-

tated from former disease, and therefore not so predisposed
to take epidemics; from the cubic space in their dwellings
being greater than that which falls to the lot of the private
soldier; and from their earlier application for medical aid on
the first symptoms of the disease.

Since the foregoing remarks were written—in fact, after
the manuscript of this work had been forwarded to England
for publication—the following admirable general order was
published by his Excellency Sir Hugh Rose, and which at
once demonstrates that the era of sanitary improvement has
commenced in India. The strict application of these regulations must indissolubly connect the name of one of the
greatest soldiers of the age with another brilliant victory,
over an enemy not less deadly, but infinitely more appalling,
than either the suns or battles of the ever-memorable Central
India Campaign; and resulting, moreover, in a diminished
mortality, perhaps even an exemption, from the fell disease
against which the directions are framed :

"Head-quarters, Simla, 7th April.

"Officers commanding divisions, stations, &c., will make
themselves thoroughly acquainted with the ground in the
neighbourhood of their stations to the extent of twenty miles,
with a view to at once selecting sites for encampments in the
event of cholera appearing, and care will be taken to ensure
these places being always kept in a fit state for occupation by
troops, and with a sufficient supply of wholesome water available on each.

"The officers of the Quarter-Master General's department
of each division will prepare a plan of the required extent of
country, with the different encamping grounds marked on it,
so that when the disease approaches, measures may be at
once taken to place the troops under canvas without delay.

"On the outbreak of cholera in an epidemic form, either
in neighbouring villages or cantonments, officers commanding

stations will be prepared to move the troops into the selected camps on the shortest notice.

" As soon as any case of cholera is reported in the station staff, the troops will be moved into camp, and no unfavorable condition of the weather is to prevent this movement being carried out.

" The force will be broken up into as many detachments as the number of the medical officers will admit, allowing *one to each party*. Should the medical staff be insufficient to afford such medical aid to the several detachments, experienced medical subordinates will be placed in charge of the smaller or less distant parties.

" Officers commanding stations are authorised to call directly for aid from other stations, divisions, or districts free from cholera.

" The sick labouring under other diseases than cholera will move with the force and share the benefit of removal from the choleraic atmosphere.

" It must be insisted on that all discharges from the stomach and bowels of cholera patients be instantly removed and buried in pits.

" Strong deodorants are to be thrown into the receiving vessels, as well as into the pits, latrines, and privies.

" Should cholera follow the troops, they will be moved short distances at right angles, if possible, to the prevalent wind and track of the disease, every second or third day, care being taken that the marches in no way fatigue the men.

" The breaking out of cholera in a regiment or at a station is, on no account, to cause the supension of the soldiers' daily amusements and occupations, care being taken that the latter in no way fatigue them ; and commanding officers will use their utmost exertions to develop any recreation or employment of which the effect is to keep the men's minds in their normal state.

"It often occurs that soldiers on a visitation of cholera indulge in the use of spirituous liquors, in the belief that they are a preventive against the disease. The medical authorities unanimously condemn this supposed remedy as a certain promoter of the disease : commanding officers are therefore enjoined to use their utmost endeavours to prevent so baneful a practice.

"One of the several cholera antidotes is the early treatment of premonitory symptoms, of which looseness of the bowels is a principal one. Commanding officers are therefore requested to give the most precise orders on the subject, and to cause all men affected by premonitory symptoms to be placed at once in a premonitory ward.

"The troops are not to return to cantonments until all traces of the cholera shall have disappeared from the neighbourhood, either amongst the European or native population. The barracks and hospitals will be thoroughly fumigated, the walls whitewashed, and the doors and window-frames painted, before they are re-occupied.

"The men will be supplied with hot tea and coffee before going out in the morning; they will invariably wear flannel belts, and all precautions must be taken to prevent their remaining in wet or damp clothes.

"The Commander-in-Chief feels persuaded that all officers share his feeling that, when cholera breaks out in a station, they should be with their regiments, and at their posts."

The sanitary measures promulgated in this order are those already recommended in this work, and some of which I have elsewhere urged at a former period.[1] It is, however, evident from preceding remarks that different soils have a greater liability to peculiar diseases than others of an opposite nature. Hence, in fixing the sites for the temporary encamping grounds, those localities should be chosen, if possible, which

[1] The Author's 'Manual of the Diseases of India.'

may happen to afford a surface of that soil which research and experience has shown to be least liable to endemic disease ; and such a clause, I venture to think, might well form an addition to the important order just quoted. Thus rocky formations generally have already been mentioned as instances of localities where cholera seldom prevails epidemically, and the probable explanation of the fact also adverted to. Other soils, however, appear to have the property of oxidising noxious germs in a quicker and more efficient manner than neighbouring surfaces. The rarity of cholera on laterite formations has latterly been prominently referred to, and cannot admit of doubt.[1]

Again, on the other hand, we know that clayey soils and most alluvial tracts retain for a long period the exciting causes of cholera and other diseases, of which the following are forcible instances :

Some few years ago a party of prisoners were employed making a road in the Guntoor district, and in cutting away the soil came upon a number of remains of persons who had died of cholera in the famine year of 1838. The disease broke out with great violence amongst these workmen.[2]

A number of coolies, employed on railway works in the neighbourhood of Salem, in cutting through an old burial ground, came upon a spring of apparently pure water. Many who drank of this water were seized a few hours afterwards with cholera of a very severe type.[3]

It is, however, unnecessary to quote continued examples of this nature. Sufficient has been stated to render it imperative that what we do know on this subject should be

---

[1] Balfour, ' On the Localities in India Exempt from Cholera.' Chevers, " On the Means of Preserving the Health of Europeans in India;" ' Ind. An. Med. Sci.,' vol. xi. Day, " On the Medical Topography of Cochin ;" ' Madras Journ. Med. Sci.,' Oct., 1861, Jan., 1862.

[2] ' Madras Journal Med. Science,' Jan., 1862 ; Review on Dr. Carter's " Fungus Disease of India."

[3] Op. cit.

practically acted upon ; also to demonstrate the urgent necessity of a strict investigation into the geological characteristics of the surfaces on which cholera occurs, or on which
it is more or less unknown in India. An enlarged and com-
prehensive inquiry of this nature cannot be thoroughly performed as a self-imposed and voluntary task ; but it appears a
subject well demanding the exclusive attention of one or
more medical officers in each presidency, who should be employed in the investigation as an especial and only duty.

An inquiry of the kind was instituted some years since by
Dr. Balfour, of Madras, and the result published in his
brochure, entitled ' Localities in India Exempt from Cholera.'
The exigencies of the service, however, caused this gentleman
to relinquish the pursuit of this subject for a considerable
period, but he has again latterly turned his attention to the
matter, and, aided by the kind offices of the editors of the
' Madras Quarterly Journal of Medical Science,' will probably again advance our knowledge in the required direction.
Unless, however, the conductor of such an investigation is
empowered to call for information, and allowed time, opportunity, and means for visiting and examination of different
localities, but little progress can be expected to be made ;
and, moreover, it would appear physically impossible that
the exertions of one individual would suffice for the efficient
examination of all the localities and of their histories which
would be necessitated in a country so vast as British India.

CHAPTER XVIII.

ON THE SOLDIER'S DIET.

THE first use of food in the animal economy is the nutrition of the body, and the second is to maintain the animal heat at a certain standard. The nutritious or flesh-forming food consists of the four elements, carbon, hydrogen, oxygen, and nitrogen; and no food which does not contain the latter can assist in the formation of any tissue excepting fat. The non-nitrogenous, carbonaceous, or non-flesh-forming food, consists only of the three elements first named, and can only be of use in adding to the animal heat or in the growth of adipose tissue.

Food, however, is generally divided into three classes, aqueous, nitrogenous, and carbonaceous.

The first comprises water and watery fluids, and is necessary to dissolve the solid food, as referred to in the chapter on " Water."

The second comprises meat, flesh, and fibrine, the white of egg or albumen, the nutritive principle of the cerealia, as wheat or gluten, and the substance called caseine, which exists in milk and some vegetable seeds, as peas and beans. Also the principle called gelatine, forming a portion of bones, skin, and cartilage.

Of the third class, or carbonaceous food, starch and fat are the simple forms, but sugar and oil belong to the same category. Without a due admixture of these different forms of food man will not thrive; one kind being taken away altogether, he cannot live.

Most of the solid matters eaten as food are, however, compound in their nature—that is, they contain both the aqueous, nitrogenous, and carbonaceous elements; as instances may be mentioned, meat, bread, potatoes, milk, and eggs.

Meat is one of the most nutritive compound foods. Beef and mutton contain about half their weight of water, the remainder consisting of nitrogenous compounds and a variable proportion of fat. Roughly speaking, in 100 parts of meat, 40 will represent the nitrogenous element, 10 the carbonaceous, and 50 the aqueous.

The average composition of bread made from wheat flour is, 70 of starch and sugar, 15 of gluten or albumen, and 15 of water; the gluten being the nutritive, the former the heat-producing food; so that in bread the properties of meat are nearly reversed. This, however, is nearer the proportion in which the different elements are required to support life.

Potatoes are less nutritious than bread, and contain more heat-forming food than meat. This vegetable contains 75 per cent. water, 23 starchy material, and about 2 albumen. Potatoes, however, are very necessary in tropical climates, for their anti-scorbutic properties, and, where green vegetables of the natural order, *Cruciferæ*, are not obtainable, should always form a portion of the diet.

It has been proved by experiment that an adult man requires daily about five ounces of flesh-forming food, and about thirty-five ounces of heat-giving food, to retain his health and strength in temperate climates. In the arctic regions, however, more of the latter is required, and in the tropics a greater proportion of the former. Hence may be calculated the quantity of meat, bread, and potatoes necessary to support life in different climates. For the five ounces of nitrogenous matter will be wanted about twelve ounces of meat. To maintain the animal heat at its constant standard, a larger amount of carbonaceous material is required. By the oxygen which enters the blood from the air breathed by the lungs, this carbon is slowly consumed or burned, forming the carbonic acid expired, and producing the internal heat. To maintain this animal heat, from eight to ten ounces of carbon must be taken in the food daily. Thus the dietary found necessary for persons fed at the public expense will be some ten or twelve ounces of meat, and about twenty-five ounces of flour daily.

Of bread alone, fifty ounces, or about three pounds would be necessary to sustain life. Of potatoes, a man must consume twenty-one pounds daily for the same purpose; of carrots, about thirty-one and a quarter pounds; of turnips, eighteen pounds, and of cabbages somewhat less; of milk, about six pounds will suffice.

Peas and beans, however, contain almost twice as much nutritive matter as wheaten flour; and an excess of the latter, but to a smaller extent, is also found in "dhal," and several other of the Indian corns.

On these facts, combined with the results of experience, have been founded the dietary scales of gaols, hospitals, and workhouses. But for men in health is allowed something more than is absolutely necessary, exercise and the consequent wear and tear of the system rendering this alteration imperative.

The rations of European troops in India are daily per man—

Meat—beef or mutton	.	.	.	1 pound.
Vegetables	.	.	.	1 „
Bread	.	.	.	1 „
Rice .	.	.	.	4 oz.
Sugar	.	.	.	2¼ „
Tea or coffee	.	.	.	⁵⁄₇ „
Salt .	.	.	.	1 „

At many stations, particularly where fresh beef cannot be obtained, salt beef or pork is issued once during the week.

Several medical men of great experience have entertained the opinion that the amount of food consumed by the European soldier in India is excessive, and that thereby a considerable amount of disease, or tendency to disease, is generated.

Dr. Chevers observes, that "there is reason for believing that European soldiers, when not on service, are rather over than under fed."

The reports of numerous medical officers tend to show a very prevalent opinion, that the quantity of meat may be reduced during the hot weather at least.

Mr. Macnamara states, that out of twenty-four cases twenty-three were found to have the liver, kidneys, or some other organ in a more or less marked state of fatty degeneration, and this fact he seizes upon as accounting for the high rate of mortality among European troops in Bengal. In the earlier stages these organs were infiltrated with fatty matter; in the latter stages this infiltration had encroached upon the normal tissues, so as to constitute degeneration.

This state of things is referred by Mr. Macnamara to the mode of living, and not to the climate. "The company's allowance is to each man per diem one pound of bread, one of beef, a quarter of a pound of rice, and half a pound of vegetables; and to this over-liberal allowance the men add themselves one pound of meat (often the common bazaar

pork ; and the meat they buy is always of an inferior quality,
as they pay no more than at the rate of two rupees for a
sheep), and one pound of vegetables, and often rice also. So
that each man consumes, on an average, seventy-six ounces
of food per diem ; whereas a man in the Royal Navy, during
the same period of time, eats but thirty-five ounces, and yet
he is in a service that compels him to take a large amount
of muscular exercise in the open air, while the man in the
1st European Bengal Fusiliers is in a situation which, unless
he be on active service, requires him to make but little
muscular exertion ; and he is, moreover, exposed to the influences of a climate which, from its very nature, would compel
him, if he wishes to preserve his health, to live in the most
moderate way possible."

Then the habits are most unsuited to persons who have too
much to eat. "After sleeping through the night in the very
hot, close air of the barracks, he rises at gun-fire, and goes to
parade, after which he commonly employs himself in clean-
ing his accoutrements till breakfast time, eight o'clock ; this
meal over, he lies down on his couch, and sleeps till dinner
time, and after dinner he generally retires to his bed again,
and sleeps, more or less, till five o'clock, the temperature of
the barracks being frequently as high as 104° at this period
of the day. I mention this particularly, because it is after
taking food that the lungs are most active ; if therefore heat
and sleep diminish their activity, we see at once how small
an amount of carbon will be exhaled by them at the very
time when their functions should be less impeded than at
any other time. About five o'clock the private has to prepare
himself for parade ; this over, he saunters about till half-past
nine, and then turns in for the night. Of course there are a
few exceptions to this—some half dozen or dozen men taking
active exercise every day—and these few are invariably found
to be the most healthy men in the regiment."

Again, the fact that a great part of the day is spent in

darkened barracks, and that spirits are drunk with considerable freedom, must also favour greatly this disposition to fatty degeneration.

Dr. Chevers points out the similarity of the morbid appearances described by Mr. Macnamara, with those which the investigations of Mr. Gant,[1] demonstrate as existing in the diseased muscular substance of Baker Street Exhibition animals.

In my opinion it is not the government allowance of one pound of bread, one of meat, and one of vegetables, with which the men are supplied, that influences the deposition of fatty material in their internal organs. It is the addition which the men themselves make to this allowance, the common bazaar pork, the eggs, the ducks, and the hot stews and curries, with the injurious spirituous liquors they obtain which induces fatty degeneration.

Lobstein, Dr. Garrod, Carpenter, Mr. Canton, and Dr. Gibb, have so decidedly traced the connection between the abuse of ardent spirits, and the occurrence of calcareous, atheromatous, and fatty deposits, that little doubt now remains on this subject.

At the same time, it is evident from the researches of Gant that over-feeding will induce the true steatosis; but I think, in the case of soldiers, the most probable cause of the degeneration is to be found *in the abuse of alcoholic liquors, and intemperance.*

Dr. A. Thompson states that private soldiers of the British army increase in bodily weight from seventeen up to thirty years of age, and that all soldiers suffering from disease between these periods of life decrease in weight. After the age of thirty, the weight of healthy men begins to decrease, contrary to what occurs in civil life, where men increase in weight up to the age of forty. Dr. Thompson accounts for the early decay of the soldier by loss of sleep consequent on

---

[1] 'Evil Results of Over-feeding Cattle.'

night duty, impure air of barrack-rooms, want of healthy mental excitement, and *insufficiency* in the quantity, quality, or variety of the food. This, however, refers to a colder climate, where a larger amount of carbonaceous food may be disposed of than under the heat of a tropical sun. With the soldier's own additions to the Government rations, there can be no doubt that he takes too much food in India; but it may be questioned if, considering *quality*, the latter are sufficient.

The meat supplied to the soldier in India, whether beef or mutton, is not generally of the best, or what in Indian bazaar terms is called "first sort." In the monsoon season, when all is green and the pasture rich and succulent, meat acquires something of the flavour of a poorly-fed beast at home; but when, with the return of the vertical sun and fiery blasts, the whole country becomes dry, parched, and barren, cattle searching for pasture in the jungle become tough, lean, and flavorless, and when killed and dressed, the intercostal muscles are actually transparent. It is, of course, possible to obtain fat cattle, even in the hot weather in India; but inasmuch as retaining them even in tolerable condition involves a considerable outlay in grain, or "gram" for their food, corn-fed mutton or beef is never supplied for the use of the European soldier. Indeed, as a sheep may be bought for one rupee in most parts of India, and to feed one upon "gram" increases his worth to six or seven rupees, it cannot be expected that regiments could be so supplied.

When meat is of an inferior description, it may be condemned by a committee, and the contractor is then forced to kill again.

In some districts, the religious prejudices of the Hindoo population against killing the sacred cow, render it impossible to supply the European troops with beef, who consequently are condemned to a diet of mutton during the whole year. This is not the case in very many cantonments; but, strange

to say, one of the places where the prejudices of the natives
have hitherto forbidden the slaughtering of beef, has been
chosen as a hill sanitarium for Europeans, viz., Mount Aboo,
where I now write.   The absence of varied diet constitutes a
great defect in a hill sanitarium; but the mountain being
" holy," the existing prejudices are bigoted indeed.   To
supply our invalids with variety of meat diet, and yet to spare
the feelings of the natives, I recommended that meat be
killed, or even cooked, at the foot of the mountain, and afterwards brought up for the use of the soldiers.

It is a question well worthy of consideration, if horse-flesh
might not serve as provision for soldiers who are placed in
such locality where beef is not procurable.   The idea at present is certainly not very palatable to the English mind,
although in Germany, as is well known, horse-flesh is publicly
sold in the butchers' shops.   It seems absurd that the horse
should be held in such ill-favour for food, as, like the ox, it
is herbivorous.   Indeed, as M. Baudens states, " horse-flesh
is extremely well adapted for food, for the animal is washed
and curried every day, and its flesh, if tougher than that of
the ox, is certainly not less nutritious.   In fact, it makes
first-rate soup."   Two batteries of artillery, of the division
Autemarre, when encamped at Baidar, fed upon horses,
and had no reason to regret it, as they escaped much of the
disease which devastated the allied armies in the Crimea.
On service with spare commissariat, it would be the height
of folly to reject horse-flesh as an article of sustenance; and
I for one do not see why it should not be used at other times,
when only mutton can be obtained, provided, indeed, horses
could be supplied without ruinous expenditure.

Again, I do not know of any sufficient reason why pigs
should not be fed for the use of the soldier, by which two
beneficial results would be achieved.   First, there would be
greater variety of diet, and, secondly, the men would not
care to buy inferior or diseased bazaar pork—meat obtained

from animals which, in the absence of better sanitary arrangements, act the part of scavengers, and feed chiefly on human ordure.

There cannot be a doubt that bad and unwholesome meat, especially pork, produces disease.

Drs. Heming and Nelson, of Birmingham, have incontrovertibly shown that measled pork is infected with the *Cysticerci cellulosæ*, now recognised as the imperfect condition of the *Tænia solium*, and, moreover, that the vitality of this parasite may *not* be destroyed by the process of cooking, and Dr. Hoile, of the 60th Rifles, has since traced the prevalence of tape-worm in that regiment to the practice of eating unwholesome animal food. The researches of Küchenmeister and Von Siebold are also conclusive, that different entozoa are developed in various animals, as the one already named in the pig; the *Cysticercus tenuicollis* in the goat and sheep, and the *C. pisiformis* in the rabbit. The same authors have also traced two of the species of *Echinococcus* that infest the human subject into a tænia that inhabits the intestines of animals.

The characteristics of good meat are as follows :—Muscles firm, fresh, and red-looking; fat white and hard; interstices between muscles and fat filled with white areolar tissue. Yellowishness, serosity, air-bubbles, flabbiness, discolored spots, granular specks, denote inferior meat.

The measle is to be detected by drawing out the animal's tongue, and pressing it firmly between the fingers and thumb. If the measle be present, it will be felt as a slight lump, and may be taken away. The same may be sought for in other muscular structures when the animal is dead.

Vegetables.—No one imagines that the vegetable ration is too great in quantity, and it frequently happens that the quality is but inferior. In many Indian stations, on account of the nature of the soil, or the saline impregnation of the water, the rearing of vegetables is a difficult matter, and soldiers are sometimes unavoidably mulcted in this respect. The potato, however, may always be procured from some other, although, perhaps, distant locality.   Such is the importance I attach to this article of diet as a means of retarding or preventing the induction of the scorbutic diathesis, that I am strongly of opinion that potatoes should always form a portion of the vegetable diet of the men.

Bread.—Indian-made bread is sometimes very good, and at others quite the reverse, and the two faults generally found with this necessary are, sourness from the formation of lactic or acetic acids, during or immediately after baking, and grittiness from the admixture of sand and dirt with the flour.

I believe that there is scarcely anything connected with the diet of our soldiers in India which might be more advantageously reformed, than the whole of our bakery management as it is carried on in many Indian stations.

Milk and bread are the two perfect articles of human food, that is to say, the articles which contain in themselves all the elements required for the support of the body.   Bread for the adult is the better of the two, because it employs the teeth, and all parts of the body to which they are a portal in the work for which they were created.   Spongy bread, since it contains about forty per cent. of water, unites meat and drink, therein having the advantage over biscuit.   It has advantage also in bulk, for the stomach was made to act upon food in bulk, and only works thoroughly when thus distended.   It has also the advantage, from its spongy structure, of presenting a large surface for the necessary action of the saliva.

The sponginess of this important article of diet is produced, as every one knows, by fermentation.   The starch in the flour is made to give off a small proportion of carbonic acid gas, and this being retained by the tenacity of the surrounding gluten, causes the mass of dough to swell up and become spongy.

To effect this purpose, yeast is used at home; in India, toddy, or a mixture of various spices, composed of saffron, "jaeful," "ellarche," "loung," and others, to the number of twelve or more. A small portion of this mixture is placed with flour for seven or eight days, and the fermenting mass caused is used to leaven the bread.

Even in the baking process in Europe the necessary decomposition of some of the essential nutritive constituents of the flour frequently results in the formation of other substances, besides carbonic acid and alcohol, as lactic, acetic, and other acids, which affect injuriously delicate digestions, frequently disagree with children, and undergo further decomposition in the stomach. In India this is the case to a much greater extent, and the bread at times is scarcely eatable.

I have frequently thought some other method of manufacturing bread might be introduced into India. That fermentation is not necessary to prepare starch and gluten for human food is evidenced from the many classes of men who prepare their "roti," entirely rejecting the process. In ancient times fermented bread was unknown. "Rome was more than five centuries old before its people learnt of the Greeks how loaves were made, and escaped from the reproach of being a pulse-eating nation."

As, however, sponginess of bread has now from habit become almost a necessity, and as undoubtedly it presents the constituents of the flour in the condition most adapted for digestion, and for obtaining the greatest amount of nutriment from the smallest quantity of material, lightness and porosity cannot be lost sight of in any attempt at improving Indian baking.

Within the last year or two, Dr. Danglish has patented a process for making "aërated bread," by which the dough becomes distended by mechanical pressure on air, and is untouched by any chemical process, and unpolluted even by the touch of a hand, the machine turning into the oven a quick succession of ready-made loaves.

It is stated of this bread that it is quite as light and spongy as fermented loaves; that it keeps better; that labouring men eat more of it, and less meat than when they use fermented bread; and that at Guy's hospital, where it was tried by way of experiment, none was left by the patients, who were in the habit of rejecting much of the bread formerly supplied.

Baking in any country is, at the least, a most unhealthy occupation, and the practice of kneading the dough with hands and feet sufficiently revolting. Bakeries at home, in the majority of instances, are underground, hot, unventilated, and undrained, lighted with gas, and fouled by the exhalations from the weary men who work in them. Bakeries in India are even still more dirty and revolting, being, generally, confined rooms, with the thermometer above 100°, dark and unventilated to a degree, and in which a gang of filthy natives knead the dough with hands and feet, and, no doubt, take little trouble to secure their cleanliness when preparing the "staff of life" for the too-often hated European.

Could Dr. Danglish's method be tried in India, I entertain sanguine expectations that it would eventually supersede the old system.

The other fault frequently met with in Indian bread, namely, the large amount of grit it contains, arises chiefly from the immense quantities of sand which float in the atmosphere during the hot winds of the Indian plains. This dust penetrates everywhere, and corn ground at such times cannot be free from sand. I am, however, disposed to think that this evil is very generally aggravated by the corn merchants or "burriyans" adulterating the stock. Be this, however, as it may, I have reason to believe that this gritty bread injures the men's teeth; that to avoid the disagreeable jarring sensation of masticating gritty matter, they frequently "bolt" their food, and thus, in a twofold manner, indigestion, diarrhœa, or even dysentery, are excited. The evil results of broken, and therefore decaying teeth, and the consequent insufficient mastication of food, are self-evident; and I think I shall not be far wrong in stating that a portion of the ill-health and deterioration of soldiers in India may be traced to injuries of the teeth, and to want of attention to those important organs. Further, I believe that the appointment of a dentist, to attend especially to the teeth of European soldiers, would be calculated to preserve the health of the men, and hence prove economical to the State. In India it is difficult even for the higher classes to procure the services of a dentist; they cannot do so without going to the presidency towns. Most medical men know little or nothing of dentistry, and hence it is impossible for the European soldier in India to obtain proper attention to his teeth. It was this feeling which induced Sir Charles Trevelyan, when Governor of Madras, to institute a dental dispensary in that city, and to require some knowledge of dentistry from medical officers. Dentistry, however, is not a subject to be acquired at once; it requires years of study and training, as the works of Mr. Tomes and others show, before an individual can be acquainted with all its branches; and it cannot be expected that medical men, amongst their multitudinous studies, can acquire the art of *mechanical* dentistry. Were an extra allowance provided for the dental charge of a regiment, it would be an inducement for medical officers to qualify themselves in this art.

The issue of drinks to the soldier will be treated of in Chapter XIX, on "Intemperance," and malt liquor recommended as the drink *par excellence*. At present all such material is supplied from Europe, although it has been demonstrated that beer of good quality may be brewed on hill ranges, as at Missouri, &c. Such beer not being so potent as the ale and porter supplied from England, would be still more suitable for the soldier's beverage. Indeed, the introduction of the light French wines would probably prove advantageous, as it is stated that among the French inhabitants of Pondicherry hepatic complaints are almost unknown, which is partly attributed to the more general use of light wines instead of beer. I do not, however, consider that malt liquor in mode-ration has any appreciable effect in inducing hepatic disease, and am disposed to consider the exemption of the French from such affections more due to their adopting a diet con-taining less animal and more vegetable material.

The evil effects of alcohol, both immediately on the struc-ture of the liver, and eventually on the latter organ and on the general system, will be commented on in the chapter on "Intemperance." Spirits, as brandy, rum, arrack, &c., con-tain from 51 to 54 per cent. of alcohol, and little solid matter. Malt liquors contain only from 4 to 8 per cent. of alcohol, and a large amount of solid material, and hence, unless taken in enormous quantities, will not induce the peculiar physiological effects of alcohol.

As William Fergusson remarked, "When the exhausted soldier is to be exposed during the night to a chilling, mala-rious atmosphere, or when he is benumbed with cold, spirits prove a sovereign cordial and support; but to administer them under a burning sun, as an article of food, or to allow him access to them as preparatory to duties of exertion or fatigue, is about as judicious as it would be to give him a blow on the head—the one would not more certainly dis-qualify him for every purpose of service than the other."

The cooking arrangements in India are susceptible of very great improvements. At the present time it is carried on by native cooks, who, to say the least, pay but little attention to cleanliness. The raw meat is daily inspected by officers, who also frequently see it as it is cooked and served on the men's table; but little is known of the treatment provisions receive *during the interval.* An "almirah" should be provided for the food of each mess, and not only the rations, but the cook-rooms and "bowarchies" should be inspected each day, so that some check would be placed on the manipulations of the latter, who, like the witches in Macbeth, may be frequently seen, in barrack and hospital kitchens, concocting nasty messes, of which the authorities know literally nothing.

A recent writer[1] observes :—"We were accidentally in the neighbourhood of a regimental kitchen one day, and there we saw, outside in the open air, a large quantity of rice, which the cooks had put into baskets, in order that the water might drain off. This rice was covered all over with a black substance, which we ascertained, on the basket being lifted, to consist of myriads of flies. We then entered the cook-room—swarming with flies—and witnessed the filthy manner in which the meat was taken out of the pots and pitched about, as if it were food for dogs."

Something of this must always be the case in the Indian plains, where flies abound in millions, and even in the palace attack the food as it is conveyed from the plate to the mouth! and where the heat of the climate renders it necessary to employ natives as cooks; but, by a strict system of inspection, much of these evils may be remedied.

That flies frequently convey the germs of disease is evident to every one who has witnessed the poorer natives affected with ophthalmia in many parts of the East, their eyes discharging matter, on which clusters of flies feast, departing therefrom to attack the healthy eye of the unfortunate person who happens to pass. Nay, even in cold climates the etiology of disease is thus traced. The 'Journal du Loiret' states :—"A dead dog was lately thrown into a ditch, in the parish of Cortal, and there left. The carcass was soon covered with flies, which then spread over the whole place. An epidemic of anthrax was the consequence."

Remembering what has been stated with regard to the

[1] 'Lahore Chronicle,' Sept. 14th, 1861.

14

fæces of cholera patients exciting that disease when taken into the healthy stomach; knowing that the evacuations of dysentery are equally dangerous; that ophthalmia may be excited by ophthalmic discharge; and that, as Dr. Budd[1] remarks, the contagious element of typhus is principally contained in the discharges from the diseased intestines, it is not too much to recommend that it should be *the especial and total duty of some one* to keep the provisions of Europeans free from contamination by flies, which contamination may have been directly brought from the rice-water stools of cholera, the slimy discharges of dysentery, the putrid evacuations of fever, the pus of ophthalmia, or from any other decomposing abomination.

It has happened before this that the men's food has been brought to table, not in crockery, but on common earthen platters, of the material of which water "goglets" are made, and which, on account of their porosity, it is impossible to keep clean. They become saturated with each day's use, and, in time, are disagreeable both to nose and palate. Crockery, glass, and even table-cloths, should be insisted upon in every European mess, and the canteen fund should be made available for this purpose.

Sameness of diet, and the want of appetite induced by the heated and enervating climate, frequently induce the men to prepare, or cause to be prepared, hot curries and stews, so greasy with "ghee" and rancid butter, and so hot with Cayenne pepper, that it is really a matter of wonder how such mixtures are retained on the stomach. Not only do these induce dyspepsia and indigestion, engender thirst, and so lead the way to intemperance, but the hot spices appear to have a direct influence in aiding alcohol and intemperance in the induction of those chronic changes in the parenchymatous structure of the liver, from the ulterior effects of which so many suffer.

[1] 'Lancet,' Dec. 6th, 1856.

It would be hard to deny the soldier made dishes altogether, but the compounds he devours under this name require supervision, so that the hot spices and peppers may be kept within due bounds.

Tobacco.—Notwithstanding the opposition which the use of tobacco has experienced, not only in our own country, but also in others, as in Spain, where, in the year 1640, Francisco de Sevia, of Cordova, assailed it with as vigorous a counterblast as was directed against it by our own British Solomon—notwithstanding that this opposition has extended over centuries, as evidenced by the late controversy in the ' Lancet'—notwithstanding the vehemence of inconsiderate declaimers, and the moderation of a Brodie, still the smoking of tobacco has extended, and is extending, amongst all classes of society, as is evidenced by the smoking-rooms attached to the Houses of Parliament, the clubs, the tavern, and the coffee-house! and by the enormous revenue of five and a half millions sterling, which, independent of its original cost, the British nation now expends on this, perhaps the least defensible, but not the most pernicious, of its costly follies.

The reasons against the use of tobacco are, that it induces, especially when used by the young and studious, impaired appetite, relaxed muscular fibre, pallid complexion, and diminished energy; that it conduces to mental disease or insanity; that it exerts an impairing influence over the generative organs.

That the symptoms first noted do arise from *excessive* smoking is not to be doubted, but we cannot with any confidence say what amount of the narcotic will induce such a condition. It varies in different individuals, according to idiosyncracy and modes of living, exercise in the open air, and frugality in other respects.

That tobacco conduces to mental disease there is no satisfactory proof—indeed rather the reverse, as it is stated that amongst the Turks, who, by their religion, are bound to

abstain from intoxicating drinks, but not from tobacco, there
is a remarkable exemption from insanity, and I am inclined
to think that if our soldiers and sailors were deprived of its
use, of its quieting influence altogether, we should have
more instances of suicide, and, therefore, as the two generally
occur in the same proportion, more insanity in the army and
navy.

With regard to the third objection, were such the fact, it
ought to be capable of demonstration, with some degree of certainty, in a diminishing population in those countries where
the weed is most used.   But in the American States and
Canada, and the North of Europe generally, the ratio of the
increase of people is remarkably high.

If tobacco cannot be proved to be so deleterious as some
of its opponents would have us believe, there is, on the other
hand, little to be said in its favour.

In persons habituated to its use, I have sometimes thought
it acted as a prophylactic measure, when obliged to pass
through or remain in malarious or other disease-generating
localities.   When soldiers and sailors are fatigued, and cannot immediately procure food and drink, a moderate use of
tobacco will enable them to sustain these wants for a lengthened period, without the exhaustion which would otherwise
ensue.   Of this I have had practical demonstration, and have
heard men on service, particularly sailors, declare they would
prefer going without a meal, rather than deprivation of their
periodical pipe.

Again, there are periods of irritation, grief, vexation, or
irritability, which occur to all the sons of Adam, and it
is at such periods that the soothing influence of the pipe or
cheroot—"the contemplative man's recreation"—exercises
a tranquillising influence on those accustomed to it.

An attentive perusal of nearly all that has been written on
the subject, and a lengthened and close observation of many
confirmed smokers, has led me to form the opinion that the

evil or good effects resulting from the moderate use of tobacco are almost *nil;* but that the good effects predominate in those who have contracted the habit; that such good effects are not sufficient to justify advocating smoking; that when tobacco is used in *excessive* quantities, it induces various nervous symptoms, some of which have been named, but does not produce positive disease, although preparing the system for its accession; and that the same quantity of tobacco exercises very different effects on various individuals; so that what moderation consists of cannot be specified. Hence, without exclaiming with Charles Lamb,

> " The old world was sure forlorn
> When wanting thee,"

I should, on the whole, be sorry to find the use of tobacco interdicted, although ready to discourage the habit at the onset.

CHAPTER XIX.

ON INTEMPERANCE.

Alcohol renders the System prone to Disease, especially in the Tropics—Shock from Alcohol—Chronic Effects: on the Body; on the Mind—Origin of Crime—Origin of Syphilis—Prevalence of Intemperance: Native Opinions —Teetotalism condemned—Good Effects of Alcohol—Army System with regard to Drinks—Advantages of Malt Liquors—Adulterations—Allowance of Drink to Recruits—Conformation of Soldiers to Moderation—Temperance Societies.

PHYSIOLOGICAL science and experience alike instruct us, that the state of system most favorable to the invasion and development of a zymotic poison is set up by the presence in the blood of organic matter in a state of change, decomposition, or fermentation. Hence the blood of the intemperate, charged with alcohol, is the condition almost *par excellence* favorable to the attacks of the causes of zymotic disease. Pure aerated blood affords the best safeguard against the violence of an epidemic; and the most perfect system of ventilation, cleanliness, and sanitation, will fail to maintain this if, by the continous imbibition of alcohol, or by excess, the functions of the skin and lungs are interfered with, their healthy relations destroyed, and their waste products retained within the body. That this is the case, is demonstrated by the researches and experiments of Fyfe, Prout, Brocker, and others, who found that the undue use of alcoholic beverages diminish by at least one fifth the amount of carbonic acid exhaled; a presence of alcohol in the circulating fluid being always associated with a diminished per-centage of carbon in the expired air and in the exhalation from the cutaneous surface, and with diminished muscular force.

Such being the case in cold climates, the same must occur in greater intensity under a tropical sun, where a given volume of air, on account of its more rarified state, contains less oxygen to consume the carbonaceous material taken into the system.

It is well known that a large or concentrated quantity of alcohol being taken, such an immediate effect may be made on the nerves of the stomach as to give rise to *shock* of the nervous centres, and even cause the heart to cease beating, and death to ensue in a very short time. This, however, is not the deleterious effect of alcohol which occasions so much disease and suffering among Europeans in India; it is the more gradual effect of the poison, the chronic alcoholism, which demands attention here. According to the researches of Liebig, Klenke, and others, on alcohol entering the system, part is oxidized, its hydrogen entering into combination with oxygen, to make water and acetic acid, which is followed by the formation of carbonic acid. During this process animal heat is engendered, and hence alcohol is regarded as a calorifiant article of food, excess of which is certainly not generally required in a tropical climate. Such alcohol as may remain undecomposed attacks the nervous system, and, until eliminated by the secreting organs, induces its accustomed effects of exhilaration or drunkenness. If the use of alcohol is continued, these effects on the nerves constitute chronic alcoholism; excess or deprivation of the usual stimulant giving rise to attacks of delirium tremens.

With these results from alcohol—viz., a superabundance of carbonic acid in the blood, and the certain toxine nervous symptoms constituting chronic alcoholism—derangement of the nutritive process, and its consequent degenerations and dyscrasia, with diminution of muscular strength, are *always* combined. The deposition of fatty matter in the tissues, or steatoss, is also a frequent resulting effect.

It is not, however, these ulterior influences alone which are excited in the human frame by the continued indulgence in alcoholic beverages. Alcohol being taken into the stomach, and particularly at a period when the organ is empty, is absorbed *directly* into the portal circulation and liver, leading to irritation of that organ, and chronic thickening of its parenchymatous structure, the consequent pressure on nerves and vessels, and the resulting neuralgic pain or hyptalgia, and dropsical effusions and swellings.

Another form of diseased liver, and perhaps more frequently met with as the appearance of that organ in those who have led intemperate lives in tropical climates, is that of general enlargement, with fatty degeneration and destruction of hepatic cells, which, however, would appear to be the first step of the changes leading to the formation of abscess, or, escaping that, to cirrhosis.

Thus, the chronic effects of alcohol are shown to be destructive to the body of the consumer; and the evil operation of the same agent on the mind might, if necessary, be pourtrayed with even more force. The loss of memory, the impaired intellect, the miserable thoughts, the imbecility, the insanity which frequently result from chronic alcoholism, need not be dwelt on here.

In addition to this ultimate destruction of body or mind, Colonel Sykes shows that the " chief cause of crime," and therefore of punishment, is drunkenness; and it might be with safety stated that this is the evil to which most of the offences recorded in the defaulters' or the court-martial book may be traced.

Intemperance also leads to excess of other kinds, and most of the numerous cases of syphilis and its consequences which the military surgeon has to treat, are primarily contracted under the stimulation of intoxicating liquors.

Well may we re-echo, " Woe unto them that are mighty to drink wine, and men of strength to mingle strong drink! Woe unto them that rise up early in the morning, that they may follow strong drink; that continue until night, till wine inflame them!"

Although it is an undoubted fact, that during the last few years the intemperate use of strong drinks has ceased to prevail among the higher classes of society, and that "two-bottle" men are now scarcely to be found; although, since the time of Dr. Wade, in 1793—who, considering the excess then prevalent amongst soldiers and sailors in Calcutta, thought it a matter of great astonishment, not that so many should be affected by fevers, but that any should escape— intemperance has declined, still the following quotations incontestably show how deep and widely spread this national vice remains amongst the lower classes.

Thus, the military blue-book, giving the sanitary condition of the British army during 1859, shows that of every 1000 admissions into hospital, 422 diseases were the consequences of the soldiers' own irregularities—intemperance or syphilis.

A late report by Inspector-General Dr. Stovell,[1] which gives the records of the European General Hospital at Bombay for ten years, presents too clear a proof of the great and still increasing prevalence of intemperance. It was seen that out of 12,325 admissions, no less than 1146 were from delirium tremens and ebrietas, being a per-centage of 9·297, or nearly one tenth of all the admissions. With the sole exception of fever, the number is larger than that of any disease named in the paper. " And on reference to the several tables of diseases given, it will be seen that alcohol has destroyed more people in this institution than either fever, hepatitis, or diarrhœa, and nearly as many as cholera."

The influence of intemperance on sickness and mortality

[1] ' Bombay Med. and Phys. Soc. Trans.'

is shown by Waring[1] in one of his statistical tables in a
forcible manner. From the calculation referred to, it appears
that the per-centage of admissions into hospitals amongst
the intemperate is nearly double that of the temperate, that
the per-centage of deaths to strength is just about double,
and that the per-centage of deaths to admissions holds much
the same ratio.

A native newspaper speaks thus of the increasing prevalence of the practice of drinking in India :

" We have said that England is the occasion of the prevalence of wine-drinking in this country. We do not mean
to say that the mere fact of studying the English language,
or the entertaining of English ideas, creates a thirst after
wine. We do not know of any English book in which it is
stated that it is obligatory on us to drink wine, or that the
glory of man consists in his getting drunk ; on the contrary,
many English books speak of the evil effects of drunkenness.
The true reason of the prevalence of this vice is to be found
in the residence of Englishmen in the country. The multipli-
cation of gin-shops in the land is owing only to the English
people. We have read somewhere—we forget in what book—
in former days some Europeans visited a king of Bengal.
They having expressed a wish for some liquor, the king
ordered some of his menial servants to take them out of the
city, the object of which was to prevent them bringing
liquor into the city. Those days are gone by. Now you
see liquor-shops everywhere, in every village, in every street.
This year you see a liquor-shop in a place where last year
there was none. Alas! what evils is wine producing in
Bengal! Whatever may be the cause, there is no doubt of
the fact that drunkenness is on the increase in the land. To
suppress this growing evil is our bounden duty. Ye young
men of Bengal! make every exertion to exterminate this

[1] " Notes on some of the Diseases of India ;" ' Ind. An. Med. Sci.,' April,
1860.

pernicious habit.   On you depends the future improvement
of our country.   Little hope will there be of our advance-
ment, if you be not zealous in the extirpation of this detest-
able vice."

Although so strongly condemning the practice of constantly
using alcoholic beverages, I must still oppose the opposite
extreme of teetotalism.   Although there are many exceptions,
I believe, as a rule, Europeans in tropical climates require
some amount of fermented drink as part of their daily
sustenance; this is particularly the case with the old resident,
and during the hot season, when the heat takes away the
appetite, and exerts its depressing and deteriorating influence
on the system.   Where the quantity of solid food is not
sufficient in quantity to supply the waste of tissue and repair
the wear and tear of body—which it certainly is not during
the exhaustion and loss of appetite occasioned by intense
heat—certain effects arise from the imbibition of alcoholic
drinks which are not marked by the nervine results witnessed
when the agent is employed in long-continued excess; these
effects, are an additional tone and impetus imparted to the
physical powers, by which the latter becomes better able to
withstand; or to reinvigorate themselves after subjection to
the several depressing and exhausting influences, as heat and
malaria, to which they may be exposed.   Secondly, alcohol
has a marked effect in limiting the waste of the body, or in
retarding " eremacausis," and it is by these two effects—
improved physical tension, and retardation of metamorphosis
of tissue—that it proves beneficial.

In proof of the foregoing, the statistics of Dr. Waring, previously referred to, tend to show that the teetotaller suffers
far less than the drunkard, from hepatitis and cholera, but
that he is more liable to fever in its worst forms; that he
is nearly as liable to dysentery as the drunkard, and very
much more so than the temperate, but that in him this disease is less fatal than in either of the others.

Hence it appears, both from theory and experience, that the *abuse*, and *not* the proper and moderate use of fermented liquors, gives rise to their deleterious effects.

Therefore, the question arises how far this moderate use may extend? how soldiers are to be obliged to conform to such moderation? and whether the system of the army has been such as to create in the soldier a propensity for drink?

With regard to the latter question, it cannot be denied that such was formerly one of the effects of the internal economy of the army. In most cases, the boys who entered the ranks were previously of sober habits—probably drunk for the first time at the period of their enlistment, and becoming, after a few years' service, "as ardent seekers after the excitement of drink as any of their older comrades."

Some years back every European soldier in India received a daily allowance of *half a pint* of spirits; and Sir James Annesley animadverted strongly on the extreme impropriety of teaching the young soldier to drink a certain quantity of ardent spirits every morning on an empty stomach, the same quantity being served out to the youngest drummer as to the oldest soldier.

The issue of the morning dram was put a stop to in 1819, and of late years malt liquor has been regarded as one of the usual portions of the diet in cantonment.

The authorised issue to the men at the present period is, one quart of malt liquor, and one dram of spirits for each man per diem. When, as sometimes happens, the supply of malt liquor in the country is small, two drams of spirits and one pint of malt liquor are allowed.

Hence, until latterly, it would appear that the system pursued with regard to supplying the troops with liquor has been such as to establish and cherish a desire for strong drinks; but this was not the only cause which fostered the disease. "Let my men be but sober and ready for parade and other duty when required, and they may spend their spare hours in the canteen and enjoy themselves," has been the creed of too many regimental commandants; and thus idleness, or rather lack of employment, and drinking with their sergeants, barrack *ennui* and impaired health, home-sickness and distaste for military life, induce the recklessness which engenders misconduct and crime.

"They drink, repine; repine, and drink again."

That this is fast ceasing to be the case, is sufficiently evident from the late General Order of Sir Hugh Rose, calling on officers commanding divisions, to transmit reports on the means of recreation, instruction, and improvements in the useful trades of the soldiers of the army, with a view to increased improvement and systematising of such means.

The question as to what may be considered the moderate use of fermented liquors is a somewhat difficult one to answer.

Dr. Chevers remarks, " The majority of experienced medical men in India, certainly hold the opinion that it is advisable to discontinue the authorised sale of spirituous liquors to our troops." I do not, however, fully subscribe to this opinion, but would limit the sale to one dram per man, a quantity which could not induce any deleterious results, but which might frequently be the means of preventing exhaustion, of retarding metamorphosis of tissue, or of counteracting the effects of a malarious atmosphere.

Of good malt liquor, one quart, with one dram of spirits, or two quarts without spirits, might be safely allowed. Unlike spirit, malt liquor is generally taken with the meals, and therefore not applied to an empty stomach. Also, unlike spirit, it contains a large amount of solid matter on which the stomach exerts its digestive powers. Moreover, it does not appear to be absorbed directly into the portal circulation and liver, as is the case with spirit, and it also exerts a certain tonic effect.

As has already been shown, the water-drinker, in avoiding

the diseases of the intemperate, is liable to those of an opposite character. Some stimulant is *actually necessary* to most Europeans in hot weather, when the heart labours and the vital energy is exhausted; and also in the autumnal season, when the physical powers are at a minimum, and malaria most prevalent. *Of all stimulants none are so innocuous as malt liquors,* which Europeans in the tropics should be encouraged to drink in *preference to all other beverages.*

In fact so much value do I place on a due allowance of malt liquor as a means for the preservation of the health of Europeans in India—as almost a necessity of the economical existence of European life in this country—that I would not hesitate to recommend the withdrawal of all duty on this *necessary.* By cheapening malt liquor in India I believe that the man who drinks largely would use this liquor in preference to all others, and thus the consumption of wine and spirits would be lessened; while the "prudent man"— the individual who denies himself " beer " on account of the expense—would not hesitate to use the beverage, both to his own health and advantage, and therefore to the interest of the State. Of what profit can it be, that a European of any class in India, should, by denying himself malt liquor, fall into an anæmic and debilitated condition, and entail upon the State the expense of his furlough to Europe? That such occurs in India on account of the expense of malt liquor can scarcely be denied. Many individuals either cannot, or will not afford to purchase malt liquor, and their vital energy suffers in consequence, and particularly so since an increase of duty has been imposed upon malt liquor. The reproach which has been applied to Europeans in India—viz., that if they vacated the country one of the chief mementoes of their former presence would be the heap of " beer bottles " they would leave behind—is not merited. Malt liquor, I assert to be one of the necessaries of life for most Europeans in this debilitating climate, and the sooner it is less expensive, the more quickly will the chances of European life in this country be raised. In the earlier periods of European occupancy of India, bottled beer was not so common a drink as it has since become, and yet statistics prove that the average of the white man's life in the Eastern tropics, is much greater than was formerly the case, to the extent of at least one half.[1] All this improvement is not, of course, due to malt liquor; but much of it may be fairly attributed to the greater consumption of the latter, and the less amount of alcoholic material now made use of.

The question, however, of the reduction of duty on beer in India is for wiser intellects than mine to decide upon. I can only advance what I believe necessary for the *great majority* of Europeans in India—for the preservation of their health, and hence for the pecuniary benefit of the State. I may, however, be permitted to give additional publicity to the remarks of Mr. Low,[2] on the subject of the measure of imperial pints and quarts; many bottles of beer one purchases in India, more nearly approaching to the former quantity than to the latter, for which they are charged.

The issue of porter to the troops has been objected to on an idea that it tends to induce liver disease. This, however, I believe to be quite unfounded and also that porter, if sound, may be drank in India with almost as much benefit as any other malt liquor. It is, however, from its colour disguising foreign material, more likely to be adulterated than its sister liquid, and therefore all porter, and in fact beer also, shipped to India for the use of the European troops should be thoroughly and severely tested previous to leaving England.

The most common adulterations of malt liquor, especially of porter, are sugar, treacle and honey, wormwood, quassia, buckbean or Menyanthes trifoliata, aloes, cocculus indicus,

---

[1] European officers are here alluded to.

[2] Low, 'Central India during the Rebellion of 1857-8.'

St. Ignatius' bean, nux vomica, opium, tobacco, coriander, carraway, and orange peel. The detection of some of these agents is easily effected by the taste alone; the presence of others can only be demonstrated by a complete chemical process, a detail of which space will not admit of.

It is, however, a question, if the allowance named should be granted to recruits or individuals newly arrived in the country. As I remarked in my 'Manual of the Diseases of India,' during the first months of residence, a young healthy individual, particularly if inclined to plethoric habits, will do well by abstaining from all stimulants, and adopting a diet principally composed of vegetable material; for if the effects of alcohol or spirit drinking, or if the habitual indulgence in hot-spiced dishes be added to the excitement of the portal system induced by the change of climate, I feel justified in asserting that such changes will be induced in the minute structure of the liver, which, as a rule, will eventually destroy the health, happiness, and even life of the patient. I would therefore limit the allowance of drink to the young and robust soldier, for the first year, to half the amount admitted to the seasoned campaigner, and interdict spirits altogether.

How soldiers should be obliged to conform to such moderation is, however, a more difficult question; and until higher influences can be brought into operation, there is but *one* course to pursue, viz., to make the already stringent rules regarding the sale of liquor to European soldiers, still more authoritative, and to treat with the utmost rigour any native so offending. If such were known to be the orders of the " Sirkar," and such were carried out in several instances, it would soon become almost an impossibility for a European soldier to obtain liquor, otherwise than the established quantity available from the canteen. Thus, not only would the supply of liquor to the soldier be stopped, but he would be prevented injuring his health by drinking the cheap deleterious trash sold in the bazaars, which is always more or less drugged with datura, cocculus indicus, gungali, cayenne pepper, or some other narcotic or pungent agent.

It may probably be objected that other means for the prevention of intemperance than orders should be had recourse to; this I readily admit, but until time elapses to allow such means to bear fruit the European soldier who will not or cannot take care of himself should have such care exerted for him, and there is no other way of doing so in India excepting by orders and severe punishment of those disobeying such orders.

The higher means for the prevention of intemperance are reduced by Dr. Macpherson to the following heads:—1. The higher influence of improved religion and morals. 2. For lower natures, the vow and oath of temperance associations. 3. The substitution of less stimulating drinks for strong alcoholic ones. 4. The punishment of the drunkard, and reward of the sober. 5. Occupying and amusing the leisure hours of the soldier.

To these may be added increased salubrity of the air of the soldier's dwelling, and the consequent absence of that depression which invariably results from residence in a vitiated atmosphere.

The first means can only be brought into operation in the course of time, and perhaps may only now be said to be commencing. The third has already been adopted in the substitution of malt liquor for spirit. The punishment of the drunkard and reward of the sober is carried out, and the occupation and amusement of the leisure hours of the soldier, as almost daily orders demonstrate, receive the constant attention of our military chiefs.

I have already stated that I cannot advise, either from theory or experience, total abstinence for Europeans in India; hence I am unable to advocate the formation of

15

temperance societies in regiments. Indeed, the plan has already had a trial, and although in some corps it appeared to work well for a time, and intemperance and crime decreased, still it was found that the men frequently broke their oath, or, as was the case with the men of the 18th, who, when ordered to Burmah, applied to the clergyman for exemption, on the ground that they could not stand the climate without spirits.

It may have been an impression that ultimately but little benefit would be derived from temperance societies which caused the Duke of Wellington to write [1] that, " while convinced that if a system of temperance could be generally established in the army, it would be greatly to the advantage of the troops, his grace is not prepared to sacrifice the great military principle which prohibits the formation of clubs or societies of any description in the army; and therefore, although the soldiers are at liberty to enrol themselves as temperance men, all meetings or associations formed with a view to induce them to do so, or to adhere to these engagements afterwards, are to be peremptorily prohibited."

[1] ' Circular to H.M. Forces in India,' 1st October, 1845.

CHAPTER XX.

Former Ravages of Scurvy—Now Endemic in India—"Latent Scurvy"—
Causes of Latent Scurvy; Diet; Light—Prophylaxis—Reaction of Scurvy
on other Diseases—Native Remedy.

ALTHOUGH we do not now hear of those sufferings from
scurvy which in former times (indeed, so late as seventy-six
years since) caused a mortality of 125 per 1000 in the
whole naval force of Great Britain, although the causes of
scurvy are so well understood, and its prevention a *certain*
result of proper prophylactic measures, still the scorbutic
diathesis may be said to be a very prevalent condition in
India, and one which the tropical surgeon is *at all times*
called upon to guard against, and very frequently to cure.

There are extensive tracts in India where scurvy may be
considered as endemic. Wherever there is a soil highly
impregnated with saline matter, especially on the borders
of the desert regions, and on much of the arid and sandy sea-
coasts; wherever, from such causes, there is an insufficiency
of nourishing and vegetable food, especially if such localities
are also low-lying, damp, and malarious, *there scurvy is
endemic,* and shows itself in one or other of the manifestations forming the scorbutic diathesis. Thus the disease
known as Beri Beri, so frequently existing on the sea-coasts,
the Delhi boil, the Scinde boil, the Gwalior ulcer, the Aden
boil, the Sunah boil, the Burmah boil, the Kutch boil, may

all be most frequently easily traced to the influence of the above-mentioned condition.

The probability of the existence of this scorbutic taint must, therefore, never be lost sight of, forming, as it does, one of the most dangerous constitutional defects, and aiding the heat and malaria of the climate in rendering the system an easy prey to acute or chronic endemic disease. Just as from the influence of a malarious atmosphere the conditions known as cachexia loci and masked malarious fever may arise, in like manner, from want of essential articles of diet, that state of system which I venture to designate "latent scurvy" originates. This latent scorbutic condition once present, influences all diseases which may affect the individual, aggravates with deadly complications all surgical cases, and obscures and masks those ailments which belong to the province of physic. Thus, ulcers take on a phagedenic character, syphilis becomes increased in virulence, and injuries, as fractured bones, will not heal, while dysentery terminates with sloughing and hæmorrhage, hepatic abscess or engorgement of the spleen is more liable to make its appearance, and the system, deprived of that vital force so necessary for the battle with disease, has little stamina to resist the latter.

As one evidence among many showing the importance of the Indian military surgeon always holding the probability in mind of the existence of the scorbutic diathesis, not only amongst Europeans, but with natives also, I will quote Dr. Steward's,[1] 'Annual Report of the 25th Regiment, N. I., for 1858-59.'

After the storming of Jhansi and the hard work consequent thereupon all the sepoys became scorbutic, and the character of the wounds was so changed that " medical officers, on visiting the hospital, remarked that, had they been ignorant of the patients having been wounded, they would

[1] ' Bombay Med. and Phy. Soc. Transactions,' 1859, p. 12.

have deemed the cases malignant fungus. The diet of the sepoy is of a nature to aggravate the above state, and we have thus presented the disease of deprivation, of low diet, and of impure oxygenation of blood, the latter aggravated by the crowded state of the hospital."

It is not, however, the ulterior consequences of scurvy to which I would direct attention in this place. Such conditions require the *curative powers* of medical knowledge. It is the prevention of scurvy which more particularly falls within the province of sanitation, the rendering impossible that peculiar condition of the system formerly described as the scorbutic diathesis, a slighter and less apparent degree of which constitutes " latent scurvy.''

I believe latent scurvy to be a more prevalent condition in India than is generally supposed. As a rule, even individuals of the better classes insensibly acquire a habit of eating less vegetable material as part of their daily diet than they consume in Europe; this partly arises from the fact of a scarcity of vegetables frequently occurring (particularly in former years) at mofussil stations, and partly because, from loss of appetite during the hot weather, so large a mass of food is not taken as in a colder climate, and the greater part of such material consists in soups, curries, beer, and other matters tempting to the appetite, and which can be consumed without any expenditure of physical force. Every one who has been long in India must admit the truth of the foregoing, and well remember that during the hot season broths and liquid food could be taken when the same in a solid mass would be loathed; and the latter being generally composed of animal substance, a diminution of vegetable constituents in the diet results.

But although the want of vegetable matter or the elements of its composition in the diet will induce scurvy, still this is *not* the sole cause of the affection. Recent experience leads to the belief that *all insufficient, exclusive, or artificial diet,* if long persevered in, will induce symptoms of the disease, but more particularly and in a shorter space of time if the defects of food involve a loss of the just proportion of succulent vegetables with their salts of potash and azotized material. Scurvy was formerly supposed to depend entirely on the use of salted provisions, but this idea is no longer tenable. It is the *absence* of certain articles of food, and not the *presence* of others, which excites the diathesis, and it has occurred that the *addition* of a small portion of salted meat to a diet composed of insufficiently *varied* materials has been followed by beneficial results. This, I believe, occurred during the Caffre war among the troops, then under the charge of Dr. Hadaway, the present Deputy Inspector-General of British troops at Bombay.

Another predisposing cause of scurvy which is in existence in India consists in the darkened dwellings in which many persons exist during the hot weather. The hot wind, and the light also, are shut out for the sake of coolness and to avoid the fiery blasts by the European; and the native, as a rule, lives in a hut with only one small door, and no other external opening. That absence of sunlight is a strong predisposing cause of scurvy we have abundant evidence in the writings of Kane, M'Clintock, and other Arctic travellers, and there is every reason to suppose that the same result occurs in tropical regions.

The prophylaxis, therefore, of scurvy consists in the *due admixture* of vegetable and animal food in the diet, and in cases where the former cannot be obtained, the use of lemon juice, tamarinds, citric acid, or potash, must be substituted. Dr. Garrod has latterly shown that the vegetables, as the Cruciferæ, potatoes, &c., which possess a reputation as anti-scorbutics, owe their good effects to the potash they contain, which, indeed, was suspected long ago by Dr. Henderson, who cured his patients on board ship by giving them potash obtained from the gunpowder.

As the scorbutic diathesis will allow the system to become an easy prey to almost any other disease, and as it aggravates all such subsequently contracted maladies, so every other disease, as malarious fever, dysentery, syphilis, in fact, every ill to which human flesh is heir, will be rendered more mortal, intractable, and virulent, by the effects of deficient or improper diet.

In conclusion, it may be remarked that Dr. M'Nab[1] has recommended the shrub called annotta, and which is indigenous in some parts of India, as an anti-scorbutic of considerable power, and one in which the natives place great confidence.

[1] 'Calcutta Med.-Phy. Society's Journ.,' No. 3.

CHAPTER XXI.

ON PROPHYLACTIC MEDICINE.

Results from Prophylactic Medicine in the Navy—Same Plan recommended
for Soldiers, in Malarious Seasons, during Cholera—Respirators.

It has long been a standing rule in the navy, enjoined by
Art. 9 of the ' Surgeons' Instructions,' that when men are sent
on shore in tropical climates the surgeon is to recommend
for each man, previous to leaving the ship, quinine and wine,
and the same on return, if necessary. Dr. Bryson ('Med. Times
and Gazette,' January, 1854) gives the reports of twelve
medical officers, showing the good effects of the prophylactic
on the South African station, the ratio of deaths falling
during the year from an enormous figure to 6·9 per 1000.

I can also bear witness to the power of quinine as a
preservative against the fevers of the Persian Gulf, as exemplified by the exemption of seamen taking the alkaloid
from malarious disease.

Hence I would recommend that, on all occasions when
soldiers are exposed to malarious influence or obliged to be
located in known miasmatous localities, and during the
malarious months, that the preservative effects of quinine
wine should be taken advantage of, by administering it to
the men.

If it be argued that the British soldier cannot be forced
to take medicine before he is sick, I would reply that any
body of men, by a judicious mixture of advice, explanation,
and jocularity, may be induced to pursue a method entailing

so little inconvenience. Call the quinine wine "a drain o' bitters," tell the men it will give them an appetite for their breakfast, or something of that kind, and there will be few unwilling drinkers. Even the latter would be overcome by the force of example, and in order to effect this the officers, medical and military, might take the same quantity. Quinine in moderation never injured any person yet, and a daily dose during the malarious season of the year, when in miasmatous localities, will often prevent that attack of fever, or delay that *cachexia loci*, under which so many, both officers and men, succumb.

Prophylactic medicine is not, however, only necessary during malarious seasons. The system may, and should be, pursued during epidemics of other diseases—cholera, for instance. Deputy Inspector-General Murray, in his 'Report on Epidemic Cholera in Central India in 1860,' ascribes the greatest importance to the exhibition of appropiate remedies at an early period in this disease, and states that a cholera "pill was ordered to be given in barracks to every man whose bowels were open during the night."

Such measures as these, as circumstances may require, should never be neglected when there is reason to suppose the epidemic constitution of the air, or the causes of disease, are more powerful than usual.

Dr. Stenhouse observes, "If our soldiers and sailors were furnished with charcoal respirators, and if the floors and tents and the lower decks of ships were covered with a thin layer of charcoal, I think there would be little to apprehend from cholera, yellow fever, and similar diseases, by which our forces have been decimated," and suggests, to avoid inconvenience, the charcoal might be covered with canvas. ('On the Disinfecting Properties of Charcoal.')

The properties of charcoal as a disinfectant have already been referred to (Chap. XV, "On Conservancy"), and there does not appear any reason why, when epidemics are dreaded or present, that this inodorous substance should not be used as recommended ; in other seasons it would be sufficient to place it on open trays or shallow boxes ; but when epidemic disease is present all consideration of appearance should be foregone, and the immense absorbent and decomposing powers of charcoal made available to their fullest extent.

In like manner charcoal respirators should form a part of the stores of every body of men, to be used when obliged to pass through known unhealthy localities. The best kind of respirator can be easily manufactured by enclosing a layer of charcoal in cotton, the power of the latter of which in preventing the passage of impurities in the air through its texture is referred to in Chapter II, " On Zymotic Disease."

Natives are well aware of the preservative effects of respiring through an artificial texture, and hence, as a rule, wrap their heads up in their clothes, particularly when exposed to or sleeping in the night air. There can be no doubt they frequently escape malarious disease by this precaution.

During epidemics of cholera, and, in fact, on all occasions when zymotic diseases, as cholera, contagious or typhoid fevers, yellow fever, &c., occur, the recommendation of Dr. Budd [1] should be attended to, and the vitality of the fæces of these ground-infecting diseases be destroyed, " by placing two ounces of caustic solution of chloride of zinc in the night-stool on each occasion before it is used by the patient," which will entirely destroy the contagious properties. Also, during the epidemics of cholera, these neutralized fæces should be immediately taken away and buried in a deep trench, as suggested by Dr. Bidie. [2]

[1] ' The Lancet,' March, 1860 ; Budd, " On Typhoid Fever."
[2] ' Mad. Quart. Journ. Med. Sci.,' No. 1 ; Bidie, " On Etiology of Cholera.

CHAPTER XXII.

ON SYPHILIS.

IT has long since been remarked that syphilis has increased in severity on armies being subjected to certain conditions, as on the march into foreign countries, during the occupation of conquered districts, and after the return of the forces to their own country or into cantonments. Thus, Fergusson informs us that more men were mutilated by venereal during the first four years the English army occupied Portugal than the register of all the hospitals of England could produce in the last century. Baron Larry remarked that after the battle of Jena the most serious malady was syphilis, and Chevers states that this was exemplified on the return of the troops into cantonments on the termination of the late Indian mutinies; at the present time the amount of venereal disease existing among the invading American armies is stated to be enormous.[1]

The causes of this virulence and prevalence of syphilis at such periods are to be found in exposure to the inclemency of the weather, want, intemperance, filth, the establishment

---

[1] 'Medical Times,' Jan. 1st, 1862.

of the scorbutic diathesis and malarious degeneration, all of which are more prominent during a campaign, and in the fact of the men being supplied with more money and from their allowing evil passions to become more ascendant after the trials and hardships of active service, and thus not restraining from the intoxicating cup, and its frequent sequel among soldiers—gregarious cohabitation.

It is not, however, previous to, during, or immediately after, periods of war, that syphilis alone induces serious loss of life, and perhaps more invaliding amongst our European troops; at all times this fell disease, either in its primary form furnishes the hospital with patients, or in its secondary or tertiary developments, *per se,* or combined with other tropical and malarious maladies, aids vigorously in destroying the wretched victim so impregnated.

Thus, from the most authentic data, it is stated that in the three presidencies the invaliding for syphilis alone amounts to 112 per 1000 of those invalided.

But the number invalided for debility and rheumatism amounts to 120·6 per 1000, and, as Dr. Lownds so justly and forcibly points out, "many of these cases of so-called debility and rheumatism have their origin in syphilis;" they are, in fact, cases of syphilitic cachexia and syphilitic rheumatism, aggravated by malarial degeneration.

In a table furnished by Sir A. Tulloch, given in the report of the late Lord Herbert's commission for the period subsequent to 1837, the rate of admission into hospital per 1000 of strength were, in Madras 315, in Bombay 262, for syphilis, annually.

Hence, from deaths, invaliding, and loss of service while men are in hospital, venereal disease plays no inconsiderable part among the maladies which swell the expenses attendant on the British occupation of India.

A recent writer has thus sketched the career of the young soldier affected with syphilis:—" He has not worn the Queen's livery for many days when he falls before allurements which encounter him just outside all barracks; he comes out of hospital with the chancre cured, but not by any means so much colour in his cheeks as when the disease was first incurred. Before very long he presents himself again, with pain in his bones, the syphilitic papulæ, and the syphilitic physiognomy, more or less strongly marked; after a certain quantity of hydriodate of potash has been swallowed, and after a few hot baths, &c., he is a second time discharged, cured. 'Syphilis' may, however, be still read in his face, and his knapsack is too heavy for him; he thinks that once relieved of that he would be all right and well again. He goes to hospital; the next discharge is a final one. The poor unit has gone to swell the terrible statistics about which we have all read, and over which it would be well for us to ponder more and more."

It is a very general idea amongst soldiers, and, indeed, amongst those who occupy a higher position in the social scale, that venereal disease, especially in its primary form, is easily and quickly cured. The sooner all people are un-deceived on this point the better it will be for the public health and morals, and the less likely will individuals be to incur the danger. There is no disease from which empirics and nostrum-mongers reap a more abundant harvest, and this arises in a great measure because they promise, without hesitation, to do what they well know they cannot do, namely, to perform a *cure*, which the conscientious practitioner shrinks from asserting to be in his power.

I will grant that in some constitutions secondary or tertiary symptoms may not occur, but their non-occurrence is the exception, and not the rule, after a well-marked primary Hunterian or hard chancre; if the individual on whom the latter occurs has any *taint of scrofula* in his constitution, I believe secondary affections *will always result*, whatever may be the treatment adopted. Certainly sequelæ may be delayed and modified by remedies, but the poison remains dormant in the system, and after the lapse of months or years shows itself on some exciting agency being applied, or probably without any evident cause.

Every medical man who has seen much syphilitic practice must admit the truth of these remarks. From the days of early pupilage under Langston Parker to the present period numberless instances have presented themselves which authorise my assertions.

The rise, fall, and commencing re-origin of the mercurial system ; the minor degree of power exerted over syphilis by preparations of potassium ; in short, the absence of any known specific remedy for the disease, alike demonstrate that *the fact of the incurability of the malady ought to be brought prominently forward in attempts at removing our great " Social Evil."*

The period has arrived when a false modesty should be avoided, and the subject ventilated. Mr. Solly, "so far from regarding syphilis as an evil, deems it a blessing," acting, as it does, as a restraint on the indulgence " of evil passions;"[1] and without endorsing this remark *in toto,* I would add to the fear of contracting the disease by making public the ravages it creates ; the inability of medicine to cure ; the peculiar constitutions it affects with most virulence ; the manner in which it recurs in after life; the propagation of the affection in offsprings; and all such matters connected with the subject.

Surely those who realise the picture presented by the following doggerel will be supplied with one other powerful motive for avoiding this disease:

> " A haggard spectre from the crew
> Crawls forth, and thus asserts his due :
> 'Tis I who haunt the sweetest joy,
> And in the shape of love destroy ;
> My shanks, shrunk eyes, and bony face,
> Prove my pretensions to the place."

---

[1] Debate on Mr. Acton's paper at the Royal Med. and Chir. Society.

While thus advocating the publicity of the dangers arising from venereal, both amongst the general public and the military, the officers and the soldiers, it is positively necessary that other methods for the prevention of syphilis should be introduced among the latter.

In former times great severity was resorted to by military authorities in order to keep their camps free from prostitutes. Thus even in 1643 runaway wives apprehended in camps were to be put to death, unmarried women branded and scourged; at other periods they have been ordered to be whipped, made to ride the *cheval-de-bois,* or had their left arms broken!

Years back, also, a fine used to be inflicted on soldiers or sailors, which became a perquisite of the medical officer. This plan, however, did not work satisfactorily, as the men, in order to avoid the payment of such imposition, used every means in their power to avoid confessing the ailment, and hence rendered it almost impossible to cure, check, or modify the disease.

At a later period this fine was commuted for an increased rate of hospital stoppage; this plan, however, does not appear to have answered, and was discontinued in 1820.

The system of weekly inspections, designated before the Army Sanitary Commission by Mr. Dartnell as "an odious and disgusting operation, and also ineffective," is stated not to have answered " the expectations which were formed of it."

Lock hospitals for diseased women were long since adopted in India, but, as Dr. Arnott tells us, " an outcry was raised against them, on account of their supposed demoralising tendency, and some even argued that, instead of removing, they increased the evil;" accordingly the Lock hospitals were abolished in 1830.

It would appear, however, that these institutions were not well conducted; that they were made unimportant duties, and given into the charge of the junior medical officer; that they were even frequently left to the care of native subordinates, who certainly "took advantage of their position, and were often guilty of gross.interference and imposition."

Medical officers, however, were not wanting who protested against the discontinuance of these institutions, which, as Dr. Arnott remarks, " *could* scarcely *increase* the evil." We have, however, direct evidence of their efficacy in the statistics of Waring, which incontrovertibly demonstrates that during the years immediately following the abolishment of Lock hospitals the per-centage of the admissions into hospital for venereal disease to strength *was more than doubled.*

The means which appear most likely to prevent syphilis among soldiers consist of a combination of several of the methods above enumerated, in the encouragement of marriage among the men, and in giving a higher status and position to the soldier's wife (see Chapter XXVIII, " On the Soldier's Wife," &c.) and in the elevation of the moral character of the soldier himself.

The subject of syphilis should be frequently explained to the men by the medical officer, and as at the present period and under present circumstances the probability of noncontact is *nil*, prophylactic measures, as cleanliness, should be enjoined. Periodical inspections may take place, but, as Mr. Dartnell recommends, at such times it is best to treat the soldier as a rational being, and look to his honour to come forward if any man has the venereal complaint. If after this the surgeon has reason to suppose the disease has been concealed, let the man be fined.

Also let the name of the woman be demanded, and, if diseased, let her be sent to a Lock hospital, which, of course, would entail the establishment of such institutions, their management under adequate authority, and the registration of all bazaar women, to defray the expenses of which, either in whole or part, the canteen funds, frequently ample, might be made available.

The question of marriage amongst soldiers is considered in Chap. XXVIII, "On the Soldier's Wife, &c., his Employment, and the Elevation of his Moral Character;" in Chap. XIX, "On Intemperance," and in Chap. XXIV, "On the Employment of the Spare Hours of the European Soldier in India."

Until, however, a different class of men than is at present the case form the bulk of our army, stringent regulations must be enforced against syphilis, as against intemperance and other causes of disease. On the ground of expense, prevention is better than cure, or, as most frequently occurs in this disease, modification. On the ground of morality, it certainly may be allowable to mitigate and check an evil we cannot at present avoid; hence, I should have no hesitation in recommending any known prophylactic agent, or method of prevention.

At present extensive experiments are being made on the subject of syphilization in Bergen, Scandinavia, Norway, and France, by Danielssen, Bœck, Vogt, Nelaton, and others, and a consideration of the arguments of both advocates and opponents of the system, tends to the conviction that in this, or some similar process, a powerful means of modifying or probably of arresting the progress of secondary syphilis may be found.

CHAPTER XXIII.

ON DRESS.

General Remarks on Dress—Former Dress of the Army—Immediate effects of Heat on a Man at Rest; in Motion—Improvements in Indian Dress—Full Dress—The best Tropical Head-dress—Objections to ventilating Hats—Leather the best Material—Form—Theory—The "Puggree"—The Stock—The Tunic—Trousers—Importance of a well-fitting Shoe—Flannel—Cotton—The Sock—Importance of the "Kummerbund"—Use of the Puggree as a Kummerbund—Use of the Beard.

MAN differs from all other animals in possessing no natural protection from the inclemencies of the weather; he has neither the warm furs of the lower inhabitants of the Arctic regions, nor the thick skins and sparse hair of animals inhabiting the tropics. Man alone comes "naked into the world," and is necessitated to tax his intelligence to provide himself with suitable covering—an intelligence which other animals so endowed by nature do not possess.

Thus we find that the pressure of circumstances in which the different races of men have been placed, has led each to adopt that covering most suitable to the situation and climate. The Esquimaux uses no flowing robes like the Hindoo, and the latter, from the force of circumstances, carefully avoids close-fitting heavy garments.

It is therefore perceived that the requirements to be fulfilled by dress, vary according to locality and climate, and indeed from employment and peculiar labour. The variations produced by climate have been already referred to.

The different dresses of the seaman, with his loose costume; the fisherman and pilot with their dreadnoughts, water-proof leggings, and peculiar shaped helmet; the miner with his flannel frock, are *apropos* examples of the effects of employment on clothing.

Such variations arising from the accident of peculiar employment being excepted, the principal physiological requirements to be fulfilled by dress are the following. The feet should be kept warm, not hot; the head cool, not cold; the neck should be moderately but not tightly covered; the chest should be left free for respiration; the abdomen supported; the weakness of the parietes of the latter, and the tension under muscular exertion, rendering this protection so necessary.

The fabric most suited for clothing is that having the cardinal property of a low conducting power for heat. This quality best preserves animal heat in a cold climate, and best protects against external heat in a hot climate. If to this be added the hygrometrical power of absorbing moisture, together with powers of radiation, we have the material *par excellence* for clothing. All these properties are possessed by flannel and woollen texture to a greater or less extent. A polished metallic vessel filled with hot water will be cooled more rapidly by giving it a covering of flannel or wool; and this power of radiation, together with inferior conducting qualities and absorbing properties, paradoxical as it may appear, render flannel textures at the same time warming and cooling, and, what is of still more advantage, induce a moderation and regulation of the temperature.

The next property which it is imperative to consider as influencing the effects of dress, is colour. From experiments it would appear conclusive that dark-coloured bodies become soonest and most heated on exposure to the sun—the maximum power of retaining heat belonging to black, the minimum to white.

It is also stated that when the sun's rays are absorbed by a dark surface, the heat loses its peculiar powers, ceases to be radiant, and will not induce inflammation or sun-burn. Such facts tend to the formation of the opinion, that in hot climates two colours should be used at the same time—white in the outer garments exposed to the sun's rays, dark in the inner clothing, to prevent these rays acting injuriously on the skin. In the African and Hindoo, with a dark skin, there is a strong desire for white clothing. The pure Arab horses—those descended from the twelve sires—have white hair, and the skin dark.

Thus it must be admitted that the clothing worn has much to do with health and disease, not only in the induction or the reverse of coughs, colds, bronchial and inflammatory affections, but also as regards other maladies which are not so well known by the public, or which are scarcely ever supposed to be due to ought else than specific causes. There are facts tending to prove that the body well wrapped in flannel or clad in warm clothing, so as to prevent a check to perspiration and chill from the night air, is more capable of resisting malaria, and hence of escaping ague and remittent fever than it would be without such aids. Again, thick clothing round the loins and bowels, especially in tropical climates, has many uses. It may afford security against lumbago, it may aid in preventing chills to the abdominal organs, congestion, inflammation of the bowels or of the liver and its appurtenances, or of the kidneys, and hence tends to check the formation of calculi either biliary or urinary.

But it would extend the limits of this article beyond reason were all the advantages derived from suitable clothing commented upon. Enough has been said to demonstrate how irrational it must be for men both in England or in the East and West Indies to wear the same description of clothing. The dress contrived for the one country and

climate, and suitable thereto, in the other becomes almost
intolerable, and the consequences have been serious. From
this cause the health of the soldier has been deteriorated and
his efficiency therefore diminished. Within the tropics, in
the old style of dress, when on duty he suffered unduly from
heat. When released from duty he threw off his warm
heavy clothing, and exposed his skin bathed in perspiration
to the breeze, and thereby not unfrequently immediately
contracted mortal disease.

Since that period, during the reigns of the four Georges,
when our soldiers wore pig-tails and powdered hair—which
they adjusted when preparing for parade by sitting in rows
on benches, each man arranging his comrade's back hair, and
then reversing position to perform the same office with the
pasted locks in front—since the period when gaiters were
worn, to fasten the number of buttons up the side of which
was a work of time and labour—nay, even during the last
quinquennial period, immense changes and improvements
have been effected in military costumes.

Army surgeons have not desisted from the days of Pringle
to the present time in urgently recommending the adoption
of suitable dress, as one of the most important means by
which the health of European soldiers in the tropics is to be
preserved. Notwithstanding the arguments furnished both
by common sense, daily experience, and physiological science,
it is only latterly that the dress of the British soldier in
India has been well adapted to the climate.

"In this country," Mr. Jeffrys observes, " the British
soldier on duty, surrounded by the hot atmosphere, with the
sun over his head, and the heated ground under his feet,
presents to our view the unfortunate subject of three hostile
agencies which induce their evil effects through the medium
of the skin or mucous membrane of the air passages, or
both.

" When the atmosphere is colder than the temperature of

the skin, the latter parts with internal heat both by perspiration, radiation, and convection. But when the heat of the air is above that of the skin the convection is reversed, bringing heat to the skin and putting its active powers to the double trial of having to throw off not only the animal heat, but invading atmospheric heat also." [1] Add to this direct solar rays and radiation, and reflection of the same from the arid ground, the rarified and therefore less oxygenated atmosphere; all of which increase the work of the skin and the mucous membranes, soon to end in their intense debility, and hence poisoning of the circulating fluid, and the physiologist has little difficulty in explaining the sequel for which the system has been prepared—dysentery, heat-asphyxia, sunstroke or fever, even when the individual affected is at rest.

As Jeffrys observes, "The brain through the aid of the *vis vitæ*, makes shift to endure for a time, but for the most part by shifting the oppression from itself to some distant organ, or to the whole skin of the body. The man sooner or later shudders (a skin impression) vomits and is in for cholera; or shudders and starts at more than a bayonet in his liver—the commencement of acute inflammation in that organ; or he shudders and dejects blood; acute dysentery with its long train of suffering is his lot." Also, "The skin's debility is malaria's opportunity." The mephites of fever, dysentery, and cholera, stalking over the bodies of a sleeping army which has been exposed to the sun by day, quickly scent out the enfeebled skins and divide the prey.

If such be the case with the European when at rest, how much more rapidly must similar disorders occur when the man, improperly dressed, is exposed to such deleterious agencies. Exertion being added to heat, immediately accelerates the debilitated circulation; the quickened coursing of the blood is attended by more frequent and deeper respirations—a vain endeavour that the blood may be cleansed and

[1] Jeffrys 'The British Soldier in India.'

oxygenated as it passes through the lobules of the lungs. The man straining to expand his chest, and urgently requiring the utmost freedom of circulation, finds that a tight fitting vestment or belt interferes with this, and that a rigid stock prevents the due return of the blood through the swollen veins of the head and neck. As the editor of the ' Lancet' forcibly observes, " The resulting distress and danger are not matters of surprise ; his muscles wearied, the man drops his head, the pressure on the stiff stock increases the congestion, this and a sense of suffocation force the dizzy soldier to release himself from the throttling stock, to unbutton his tunic, and to take whatever means he can to breathe more freely, to give freer play to the struggling heart. If these measures of relief fail, or if not being resorted to the man falls down dead or dying, he is said to have died of sun-stroke. No doubt the heat of the sun is a predisposing cause, but it is not the sole cause ; many cases of reported sun-stroke and apoplexy are the direct consequences of the mechanical interference with the circulation caused by the soldier's dress."

Lastly, there can be no doubt that in some instances even when a man is at rest (and much more quickly during the predisposition caused by exertion), that the heat of the sun's rays acts directly on the inadequately protected brain, and destroys the vital powers by shock, commencing with well or faintly marked collapse, and terminating rapidly by apoplectic congestion.

Happily these appalling terminations of life from improper dress are not so likely to happen in India at the present time as they were in former years. Efficiency is not now as of yore sacrificed to appearance ; and it is only when the Indian soldier is in full dress that he must still put on "the cloth coat of a northern clime, trousers fit for a Siberian campaign, and a hat that protects the head neither from heat or cold."

A good soldier's head-dress is still a desiderative for India.

Such an article should be capable of being thrown about without injury; should be composed of material not too heavy, and yet is required to be of that texture which will least absorb and conduct heat, and best reflect and radiate. As expressed in my 'Manual of the Diseases of India,' I am strongly of opinion that the majority of the head-dresses manufactured on ventilating principles *alone*, are grave mistakes. A ventilating hat with a wide circle, or holes, all round the head, through which the hot wind is constantly rushing, must tend to induce rapid evaporation from the scalp, which in a lengthened exposure may result in heat and dryness; thus actually predisposing to the affections which the hat was intended to obviate. On the other hand, a head-dress having a canopy into which the head fits, of necessity retains the perspiration and moisture, and materially lessens the chance of sun-stroke, or head affection.

I am aware that it may be objected that if perspiration lies wet on the surface, the skin begins to throw out less than is natural or necessary, and that internal heat may be prevented evaporating by keeping the surface unventilated. To this I would reply, that soldiers in India seldom wear their solar head-dresses sufficiently long to induce the scalp to acquire a *new habit*, and that removal for a moment, or simply adjusting the hat (as all persons involuntarily do at intervals), is sufficient to allow a stream of air to rush underneath which inversely with its high temperature cause the wet head—as I have hundreds of times practically experienced—to become most cool and comfortable for a considerable period afterwards.

I do not, however, disapprove of the ventilating system *in toto*, but I consider it should be combined with other principles and not alone depended upon.

Of all known materials, I think leather is the one best fitted for the manufacture of Indian head-dresses. It may be thrown about without injury; cannot readily be destroyed;

may be bent and restored to its former shape, and is not very heavy; advantages, excepting the last, which are not possessed by felt or pith, and scarcely to so great an extent by wicker work.

The form of all Indian head-dresses should be helmet-shaped, which alone affords proper protection to the back of the neck and temples. The present wicker hats now in use in many regiments having a broad, flat brim, allow the slanting beams of the morning and evening sun to pass below them, and strike the head of the wearer. Many individuals feel the horizontal rays even more than the vertical meridian sun, and hence it appears a *sine quá non*, that all head-dresses should have the curved brim.

The description of solar hat I prefer is one consisting of two layers of leather, the inner layer fitting round the head with a lining pad, &c., and leaving a short space between the vertex and the inner, central surface. The outer layer joins the inner one at that point corresponding with the commencement of the slope of the brim, and then ascends with the true helmet shape, the distance between the two layers gradually increasing from the junction to the summit where three inches or more would intervene. Of course props may be placed to keep the two layers separate if desirable, and such props might consist of leather, cork, or vulcanized India rubber. Unless, however, the helmet be subjected to very rough usage, the inherent stiffness of the leather is sufficient for this purpose. A small ventilating apparatus should be placed at the summit, and small eyelets to correspond at the junction of the two layers below.

By this helmet the head is kept moist; a stratum of non-conducting air is placed between the outer and inner canopy. Sufficient ventilation is afforded by the eyelet holes in the inner leather, or skull cap, and the moisture passing through the latter tends to retain the stratum of air between the leathers in a damp condition; which moisture the small amount of hot air rushing round the head is unable to dissipate.

A white cover may be worn over the helmet, or, on seasons of exposure, a good puggree or turban. Elwood's patent felt helmet is to be recommended, but is too expensive for soldiers' use.

The stock—so long denounced by all military, medical officers—is now no longer worn in India; a simple necktie being adopted in its stead.

Nothing can be more satisfactory than the form of the tunic now in daily use. It fits lightly to the person, hangs loosely, does not incommode or prevent agile movement, and when in length suited to the height of the wearer, looks well. I think, however, pockets in front would be an immense advantage, especially to men on the march, or on active service. By general order, May 21st, 1858, the shakee (ash-coloured cloth) was ordered to be the material for the summer tunic. This, however, when washed, loses colour and ultimately becomes white, thus causing a regiment to present a motley appearance, and I think the adoption of white structure, as drill, for the tunic would be preferable.

In the cold weather the tunic is worn of coarse red serge flannel, than which nothing can be more appropriate material.

I am of opinion that trousers are the best nether garments for the soldier; they are slipped on and off without trouble; may be rolled and tied up from the dirt and water, neither of which advantages are possessed by the many-buttoned leggings and breeches, or even " knickerbockers." American drill or dungaree is the proper structure for trousers. Soldiers cannot march in India in cloth pantaloons with any degree of comfort or convenience.

Marshall Saxe used to say that legs win battles; and another eminent general tells us, that the fortune of a campaign may be settled by the method of lacing soldiers' boots. Moreover, at close quarters, a soldier who is lame or weak in the feet, from ill-fitting boots or otherwise, can never apply with advantage the strength of his arm when wielding the bayonet or in sustaining a charge, simply because, as Dr. Camper [1] observes, in both cases "the foot is that part of the mechanical system of leverage that rests upon the fulcrum, the ground; so that if leverage is weakened at that important point, the strength of the whole system is immediately reduced—a fact well known to engineers."

Hence the importance of each soldier having his boots made to measure, and rejecting all misfits, by which his strength and usefulness will be greatly increased in the day of trial, and extra carriage for the foot sore avoided on the line of march.

Under the system of workshops now attached to all European regiments, it may be hoped that the men before long will be supplied with better fitting shoes, constructed or superintended by European workmen, few natives having any idea of fashioning a good fitting boot. Of all varieties of the latter, I believe the lace-up or "blucher" best fitted for the soldier, having, however, only three pairs of eyelets.

Some military surgeons, amongst whom is W. Fergusson, condemned the use of flannel next the skin, as "a nuisance in barrack life," chiefly in consequence of its being allowed to become saturated with dirt. Of the sanatory value, however, of this texture next the skin of the European in India, there can be little doubt, preventing as it does, much of the danger arising from sudden alternations of temperature. When flannel irritates the skin, as is occasionally the case with some individuals, silk is a good substitute. Very dark flannel which harbours dirt should be avoided for the use of soldiers, and frequent examination of white instituted to ensure cleanliness. As Sir Ranald Martin [2] points out, the

---

[1] 'On the Best Form of Shoe.'

[2] Martin 'On Tropical Climates,' p. 125.

cotton shirt from its slowness of conducting heat, is admirably adapted for the tropics. Cotton clothing is not only cooler than linen from the fact of its conducting to the body excess of external heat more slowly, but also is warmer from the same non-conducting characteristic, when the temperature sinks below the heat of the body.

Hence, when vicissitudes of temperature occur, the advantages of cotton are evident, and under the profuse perspirations of the tropics, when linen so wetted would feel cold and chill from accidental breezes, the cotton shirt, as stated, maintains a more equable warmth.

The socks issued to soldiers should be either all wool or a mixture of wool and cotton. Nothing tends to blister the feet so much as marching in cotton socks under a coarse shoe in warm weather. No part of the perspiration is absorbed by them, but the moisture remains on the feet, which, immediately exercise ceases, become cold and clammy.

Silk has been mentioned as a substitute for flannel underclothing when the latter irritates a too sensitive skin. Silk in contact with other fabrics will probably conduce to health by exciting an electrical atmosphere adjoining the person, although its powers of regulating temperature are inferior to flannel.

Mr. Stewart [1] says, " I believe the sun exercises as injurious an effect on the organic or sympathetic system through the medium of the spinal column and of the solar plexus, and on the action of the heart and circulation, and hence the oxygenating function of the lungs, as it does on the brain." The same has been observed by many other medical officers, and in fact I believe that many of the so-called cases of heat asphyxia have their origin *in intense spinal congestion.* Hence the importance of the turban, shading the upper part of the medulla spinalis, and of the "kummerbund" worn round the loins. All soldiers should be habitually required to wear the latter, in the shape of a double flannel belt eight inches

[1] Annual Report 25th Regiment N.I., Trans. Bomb. Med. Phy. Soc., 1861.

broad, round the waist and bowels, both day and night. When campaigning, in order to guard against the physiological well known evil effects of blasts of cold wind, or even heated air, they ought to be ordered to take their cotton turban from their helmet and wear it (as the native does his kummerbund) as an additional protection, during the hours of sleep and rest, to that most susceptible part the abdomen. The turban might be so worn during the early hours of the morning march, restoring it to the helmet when the sun's rays rendered it necessary to furnish additional protection to the head and upper portion of the medulla spinalis. By this means the former would be lightened of its burden during the absence of the sun, and the latter would be protected during the colds of night; and *vice versá*, just those periods when protection is required, and may be dispensed with, by the parts in question.

Of all the contrivances to fasten on the knapsack, that known as Berrington's appears the best qualified to remove undue pressure from and admit perfect expansion of the chest, and also freedom for the play of the arms. Although soldiers, as a rule, do not carry their knapsacks on the India march, still as they are occasionally required to do so, they should be supplied with that formation which can be borne with most ease.

Bichat long since asserted that although numerous causes might exist to produce debility of system in coincidence with the presence of the beard, yet the general impression must be that there exists " un certain rapport entre elle (la barbe) et les forces :" that the muscular energy is to a certain extent dependent on the presence of the beard, and that the habitual deprivation of this appendage causes a diminution of the vital and therefore physical force. The same idea is held by other authorities, among whom Dr. Holland[1] states, " Cutting

[1] 'The Constitution of the Animal Creation, as Expressed in Structural Appendages.'

the beard as it grows is not only absurd, but frequently prejudicial to the healthy condition of organs more or less interested in its development. The beard is the distinguishing appendage of man, and has generally been uncurbed in its growth in those nations which have in succession been pre-eminent for valour and influence of character. The beard uninterrupted in its growth is more intimately associated with the mental powers, affecting their manifestation, than may be supposed possible at first."

In addition to these cogent reasons for the preservation of the beard, there is no doubt that the moustache acts as a species of filter to the air, and thus tends to prevent the inspiration of organic material floating in the atmosphere.

Hence it is to be hoped that the soldier will never again be obliged to destroy his martial appearance, and deteriorate his health by the process of shaving, and that the beard will ere long become a distinctive feature of a " man-of-war's " man as it is now of a British merchant sailor.

CHAPTER XXIV.

ON THE EMPLOYMENT OF THE SPARE HOURS OF THE EUROPEAN SOLDIER IN INDIA.

Why should the Soldier deteriorate more than other Europeans in Tropical Climates—Ennui—Intemperance and Vice proceed from Ennui—Cause of failure of attempts to occupy the Soldier—Sir H. Rose's Orders—Numbers of Mechanics in a Regiment—Soldier to be Utilized—Objections refuted —Soldiers at Hill Stations—Sir H. Laurence and the Lahore Garden— Colonel Robertson's Opinions—High prices of European Labour—Remedy —Sir Hugh Rose's Opinions.

THERE exists no good reason why the soldier should deteriorate more than Europeans who are resident in a tropical climate, or why, excepting from the chances of warfare and active service, he should not at the retiring age of forty, be a better man, and a more useful member of society than when he first enlisted? Why should he, as one doomed, sink under climatic influences which the European officer or the Euro-pean, merchant brave with comparative impunity? or why, within the period that has but just sufficed to qualify the civilian for the higher posts in the State, should the soldier become only fitted for the infirmary or asylum?

One amongst other causes, which add to this mental and physical degeneration, which induces intemperance, the contraction of syphilis, and their results, has already been stated to be lack of employment—the demon " ennui."

Martinets have said, occupy the soldier with his drill and professional work; but that has now become an exploded policy. What a man does well, he does willingly; if work is not performed with inclination thereto, such business, in the majority of instances, is not satisfactorily performed. Keep the soldier constantly at drill and he becomes disgusted with it, and dissatisfied with his occupation; vary his professional work, with other suitable employment, and he recurs to both with zeal and zest. Moreover, it would be impossible, without overworking the officers of all classes, to keep the soldier at *perpetual* drill. Therefore there must be periods of enforced idleness, and there is not a question that out of such forced idleness and its consequences, arise the larger portion of vice and physical debility which a barrack life too often entails.

Many experiments hitherto made for providing occupation for the European soldier in India have failed, simply from the fact that sufficient attention has not been paid, in the first instance, to the various habits, tastes, and pursuits of different individuals, and recreation ceases to be such if forced; becoming, in the opinion of the soldier, only another kind of drill. A man cannot be made to dig in a garden, or find recreation therein, by putting a spade in his hand, and lecturing him on the value of horticultural pursuits. Thus it is with all occupations—the man who loves the one will often detest the other, and hence the soldiers should be left as much as possible to their own liberty of choice.

The enlightened orders of Sir Hugh Rose sufficiently demonstrate the importance which that gallant General attaches to the suitable occupation of the leisure hours of the European soldier in India. His Excellency has desired that commanding officers will at once take measures to establish in the barracks occupied by their regiments or batteries a system of workshops, " on the principle herewith promulgated, and that they will do their utmost to ensure success."[1]

Such principles are, to enable the European soldiers to carry on in their leisure hours the trades and occupations by

[1] General Orders, 1861.

which they earned their living before enlistment. To afford
opportunities for their children to learn trade. To make the
canteen fund available for prizes to good workmen. To place
the earnings of the soldier in the Savings' Banks. Defaulters
to be prohibited from working. Tradesmen of all kinds to
be distributed in each company. Workmen who are well-up
in their drill to be excused attending ordinary parades during
the hot season.

In a cavalry regiment only 654 strong there were 20
carpenters; 10 masons; 23 shoemakers; 10 smiths; 41 clerks;
3 carvers; 27 grooms; 12 bricklayers; 8 saddlers; 24 tailors;
2 watchmakers; 1 wheelwright; and 126 labourers, &c.

No wonder that with supplies of workmen like this, that
an executive engineer could report to His Excellency, at a
late inspection, that the men repaired the whole of the barrack
furniture; or that the Head Quarters' Wing of Her Majesty's 6th Regiment at Sinchal "have effected extensive
repairs in their barracks, which but for their assistance would
scarcely have been habitable."

It has been too much the custom to look upon the soldier
as a mere machine for the performance of military duty, and
to supply him, by native agency, with everything he requires.
I would endeavour to utilize the soldier, and make him perform
numerous matters for himself, which native servants are now
entertained to do. I may be met by the objection, that such
a line of procedure would expose the soldier to the sun and
vicissitudes of the weather, and so render him still more liable
to disease; but under adequate regulations this need not be
the case. Many trades may be carried on in-doors; and in
the cold weather, and often in the monsoon, there is nothing
to prevent an European working in the open air for several
hours both morning and evening.

At a hill station there is no reason why the soldier
should not at all times perform nearly every duty for himself:
feed, tend, and slaughter his own meat; build; cut trenches;

17

drain; make roads—in fact, be employed, under certain restrictions as to exposure, in every possible manner.

As an example of what may be done on the plains, under one head, may be instanced the garden constructed under the superintendence of Sir Henry Larwence at Lahore, where an extensive native garden, overgrown with weeds and brushwoods, was converted into a place of public resort for the amusement and instruction of the European soldiers: "where they could spend the day in the shade, or read, or play at all sorts of gymnastic games, according to their tastes."

Colonel Robertson, in his report to the Commander-in-Chief, "points to the necessity for various departments to employ soldiers in such work about the barracks as they are competent to perform;" and Government may readily demand from their servants that such work is carried on at a reasonable rate of remuneration.

A bar to the employment of Europeans by those requiring their services has hitherto been, the enormous rates of pay the former require. By utilizing the soldier more, by expecting each man to work at his trade, competition would be originated by which monopoly, the parent of extravagant prices, would be destroyed.

This subject cannot be better concluded than by again quoting Sir Hugh Rose:—"One of the chief objects of work-shops is to instruct the children of soldiers in trades. In this view commanding officers are requested to do their utmost to induce parents to allow their children to receive the necessary instruction. Remuneration will be given to the soldier who teaches the child a trade, the amount of which will depend on the degree of efficiency which he has enabled the child to attain."

CHAPTER XXV.

ON CAMPS AND MARCHING.

In all possible cases Sanitation should regulate the choice of Camp Sites—
Grounds to be Chosen and Avoided—Space between Tents insisted upon
—Conservancy arrangements—Graveyards to be Avoided—Cholera—Un-
healthy Camps—Period to March—Causes of Disease to be avoided *en route*.

ON active service military necessities will often necessitate
the formation of a camp in malarious or otherwise unhealthy
localities ; but at all other periods the site, aspect, and position of camps should be regulated by sanitary principles ;
and this is most *especially* demanded when there is any
probability that the camp will continue a *standing* one.
The worst ground, as already referred to (Chap. XI, " On the
Choice of Stations"), is a clay soil, or a clay subsoil, coming
near the surface, both of which are retentive of moisture,
and keep the locality constantly damp and in a malarial
condition. A few holes dug over the proposed site will
always show if the subsoil be dry or otherwise. Ground at
the foot of a slope is apt to be damp and unhealthy, as it
receives water from the higher levels. Localities of this
nature, occupying the angle between hill ranges and the
lower flat country, or situated in deep narrow valleys, often
predisposes its occupants, even in temperate climates, to
epidemic disease. High positions exposed to winds blowing
over low marshy ground, even if at a great distance (see

Chap. III, "On Malaria"), are sometimes unsafe on account of
fevers.   In tropical climates, camping grounds at the mouth
of narrow wooded valleys down which the wind blows, often
predispose to the same disease, and should be avoided.
Ground covered with rank vegetation is unhealthy, on account
of the amount of decaying matter in the soil, and the presence
of such vegetation is in itself indicative of a marshy subsoil.
Sandy tracts of rivers are also malarious, and the same may
be stated of their emboucherés.   Ground abounding with
frogs, and birds which inhabit marshy localities, is to be
avoided.   Generally a friable soil, not encumbered with
vegetation, with a good fall for drainage, not receiving or
retaining the water from higher ground, and open to winds
blowing over no marshy or unwholesome land, will afford
the greatest amount of protection from disease which the
climate admits of.

A recent writer observes, "Camps are not unfrequently
arranged so that the tents touch each other, except where a
narrow passage is left between the rows for access."   Such
a camp can neither be clean nor healthy, and thorough
ventilation is impossible.   Closely packed together barrack
huts or tents are a direct cause of disease, and the segregation over a large area is the best means of preventing or
arresting the spread of epidemic disease.

It was shown in the report of the royal commission on
the sanitary state of the army, that the Quartermaster-
General's instructions issued at the commencement of the
Crimean war authorized densities of population on the camp
surface equal to 347,000, 348,000, and 664,000 inhabitants
per square mile.   The lowest of these densities is double
that of the most thickly populated district in England.
Taking the ground *actually covered* by tents, the regulations
gave a density of population equal to 1,044,820 per square
mile.

In comparison with these figures the report mentioned

gave the following examples of the most densely populated districts in the metropolis :—

St. James, Westminster 144,008 inhabitants per square mile.
Holborn . . . 148,705 „ „
Strand . . . 161,536 „ „

An Indian camp, however, is spread over an immensely larger surface than is noted above. Every regiment has its large distinguishing flag, every tent its sign stuck into the place for its pole, all laid off by line and compass the evening before by the *quartermaster of corps* or other official, who goes forward for that purpose.

The choice of the locality for forming the camp should not however be left to the quartermaster or commissariat officer, *but the medical officer should always proceed in advance*, and apply his *special* knowledge to the important matter under consideration.

Great attention must be paid to the conservancy arrangements in a standing Indian camp, even although the canvass city may be only stationary for a few days. No dirt should be permitted to accumulate either from the shambles, bazaars, camp followers, or soldiers within the limits of the lines of tents. Trenches should be dug to leeward, and all ordered there to pay their devotion to Cloacina. Such trenches, Dr. Hamilton[1] states, " should be narrow and deep so as to leave as little surface for evaporation as possible, and every day a certain amount of earth excavated in their construction should be replaced until the whole is covered in, and then new sinks should be dug." Camp latrines should not be placed on running streams, as the communication of zymotic disease through the medium of water (see Chap. XVI, " On Water") has been so thoroughly proved, that no doubt now remains that such a course may be fraught with the greatest danger. Soldiers are ever ready to use running streams for disgusting

[1] ' A Practical Treatise on Military Surgery,' Boston, 1861.

purposes, and it would even be advisable to post sentries over the water to prevent such pollution. If the contaminated stream flows toward the enemy, disease may be induced in those regions over which the conquering army has to pass. If the water runs in a contrary direction, the rear, the advancing reinforcements, or by a retrograde movement the army itself, may become epidemically diseased from the unclean water. Nay more, the fluid itself, ere it leaves the camp, may convey sickness.

Care should be taken that camps are not pitched on old graveyards, or over ground charged with organic matter. The outbreak of smallpox at Quebec from the opening of an old smallpox cemetery has been already noticed in a previous chapter. When the first division of the army arrived at Varna, 13 June, 1854, they were healthy, till they encamped at Aladyn. There they unfortunately took the same site which the light division had formerly occupied, and then cholera, preceded by diarrhœa, broke out.[1]

On the 20th October, 1854, the battle of the Alma was fought, and afterwards the 4th Division encamped on ground recently occupied by the Russians ; cholera also raged in this camp.

In India numberless instances of the kind have happened, as the fact noted in the chapter on cholera, that the disease frequently follows the lines of the great roads or streams of communication, sufficiently shows. Hence the desirability of *not occupying ground which has been recently encamped on*. Such ground becomes fouled and dirty, and contains the elements which may be lighted into epidemic disease. The practice of having appointed spots of ground along the great routes as camping sites, is not to be recommended.

In the camp, or on the march, the same or even greater attention should be paid to the purity of water than has been asserted to be so necessary in cantonments.

[1] ' Parl. Papers of Sebas. Com.,' p. 74.

If obliged to camp in manifestly unhealthy localities, as at the foot of ghauts or on marshy ground, a prophylactic dose of quinine should be administered, and the floors of the tents, particularly of the hospital, raised with larger masses of straw or hay. Fires about the encampment will also tend to destroy malaria.

Much difference of opinion has arisen respecting the period of the twenty-four hours which should be fixed on as the most appropriate season for marching. Eminent authorities have not been wanting who have condemned the general custom of marching in the early hours of the morning. Thus Ballingall recommended "an hour's more sleep and an hour's more sun;" and it has also been brought forward as an argument against early marching, that the greater number of cases of insolation take place during the early hours of the morning. Other objections are, that the men are called out into the damp cold air of night, perhaps remain standing about for some time, and thus are liable to chills, and the consequent colicky pain in the bowels, or diarrhœa, or recurrence of ague. Annesley also advised that night marches should if possible be avoided, on account of their depriving the soldier of his rest, which cannot always be procured in the heat of the day ; and this author moreover states, that in the Carnatic, where damp is not so prevalent, early hours would be advisable ; while in Mysore, Hyderabad, and the ceded districts, where there are heavy fogs and dews, troops should not be taken out before such are dissipated.

As a general rule, it appears to me that the afternoon is the proper period for marching ; but in the very cold weather of Northern India, there is no reason why the troops should not move at daylight, commencing the march with the first glimpse of dawn. When marching at other seasons, four or five o'clock in the evening appears a more suitable season than the early hours of morning, the system, fortified by the previous dinner, being less liable to the poisonous effects of

malaria, than it is when sleep has been disturbed, and the
digestive organs long unemployed. The objections frequently
made to marching in the after part of the day are, that the
men arrive on the camping ground in the dark, that they
cannot see sufficiently well to pitch the tents in good order,
and that considerable confusion, and oftentimes accidents,
result. During that period of the month when the splendid
moon of the tropics affords light, all these objections fall to
the ground. At other times it must be admitted that they
have considerable force, especially when applied to large
bodies of men camping on an unknown or imperfectly known
site. When, however, the party is small, as a wing of a
regiment or battery of artillery, and when the camping
ground is an established one, and therefore well known, little
more difficulty or confusion would be experienced in forming
a camp, than is the case when the men, in the dark, or by
the light of fires, strike their tents and pack up their traps to
move off in the early hours of the morning.

Of course the cold season, or the period from November 1st
to the end of February, is the time when all troops should,
if possible, be moved. Instances are numerous where the
most disastrous results occurred to regiments, both European
and native, marching during the hot or rainy months.

It may be taken as the rule that troops, especially Europeans, cannot move in the hot weather or rains, without ex-
periencing a large amount of present sickness. Or supposing,
as, however, rarely happens, that cholera or sun-stroke do
not devastate the ranks while on route, still the after effects
are observable in the large per-centage of sickness always
occurring on the return of the troops into quarters. Extreme political necessity alone justifies the movements of
European troops during the hot and rainy seasons of India.
Humanity to the men, and economical considerations to the
state, together with the teachings of physiological science,
alike forbid the exposure of the European to the summer

heat of the tropical sun, or to the damp and dysentery of the monsoon deluge.

The chief causes of disease to be avoided on the line of march are, damp, fatigue, the sun, and malaria.

Damp will be least felt by the use of a tarpaulin sheet laid over the straw or kerby forming the bed. Where, however, lines of route and specified camping grounds are used, it would be well to supply the latter with wooden frames and tapes, or *charpais* for the soldier's bed.

Fatigue is a frequent source of disease, and hence, unless military necessities are paramount, the distance of each day's march should not be far, or the pace fast. A fourteen-mile march (farther than which should not be attempted in India) cannot be made, including halts, in less than four hours and a half. The pace should be regulated at a different speed during each half hour, and not left to the pace of the horse of the leading officer.

Nothing predisposes to fatigue more than thirst, and hence it is important that a sufficient number of " bheesties" accompany the column, to supply the men with good water. Each soldier, moreover, should be furnished with a string for his tin can, in order that, on the arrival at a well, he may draw water for himself; but such water should be previously examined by the medical officer.

Before leaving the camping ground, and during the march, each soldier should be supplied with a drink of tea, coffee, or conjee.

Too much care cannot be taken to avoid exposure to the sun ; therefore the tents require careful examination as to their capacity and thickness of texture. The men should march in helmets, with the puggree round the bowels, removing it from the latter to the head when the sun mounts above the horizon. (See Chapter XXIII, "On Dress.") On arrival at camping ground, it will be better that the tents should be pitched away, and the tent pegs not *dressed*, than that the men should be kept in the sun striking and pitching their canvass houses several times, or until all is taut and straight. Such procedure keeps the men without their breakfast, as well as exposing them to the sun at a period when, with an empty stomach, they are less able to resist its effects. It would be much better to practise the men at tent-pitching during the early period of the halting mornings; and the rule might be that those not pitching their tents properly *en route,* should be so employed when the other men are allowed their liberty.

Malaria must be guarded against by the scientific choice of the camping ground, by prophylactic medicine, and generally by those sanitary measures which are reiterated in this book with reference to this particular danger.

CHAPTER XXVI.

DR. KIRWAN, in his admirable ' Notes on the Despatch of Troops by Sea,' founded on an experience of numerous voyages in medical charge of soldiers, states that the means of conveyance now in use afford no appliances for the care of the sick ; for the prevention of the spread of infectious or contagious disease; for the drying of wet clothing in heavy weather; for the preservation of the cleanliness of the soldier and his clothes; for the segregation of married men and their families ; or for the maintenance of that continual drill which is so essential for the efficiency and soldier-like bearing of the men.

Dr. Kirwan, and other medical officers, have entertained the opinion that England should possess a fleet of Government troop ships, available for the despatch of soldiers at any moment, with sufficient space and proper appliances.

The approximative rate of mortality of adults on the voyage out to India, given by Dr. Wilson,[1] is 4·05 per 1000.

Dr. Chevers[2] states, among 23,801 officers and men who

---

[1] 'Guide to Health in Troop Ships and Emigrant Vessels.'
[2] 'Indian Annals Med. Sci.,' No. X.

have arrived from England, the Cape, &c., on sixty-six vessels, the average duration of the voyages being three months and twelve days, the deaths were only thirty, or at the rate of 4·450 per 1000 per annum.

From Dr. Cole's statistical table (in his 'Medical History of the Troops on board ship from England to Bombay and Kurrachee, which arrived during November and December, 1857, and January and February, 1858, Trans. Bomb. Med. Soc.,' No. 5), it appears that out of 9,348 troops embarked— women and children included—there were 2,190 cases of sickness, 704, or about *one-third of the whole, being of venereal origin,* and hence it is reasonable to suppose, for the most part, contracted prior to embarkation.

The ratio, therefore, of sickness on board troop ships to India contrasts favourably with that of men in barracks in any climate, even if we include the venereal admissions.

Thus the ratio of sickness per 1000 of mean strength is given in the Guards as 862. In the line, in Great Britain, as 1·044; in the Madras army as 1·634; and amongst the troops at sea as 680.

In the second number of the 'Madras Medical Journal,' a tabular statement is given, showing the number of troops arrived from England at Madras for a period of ten years, with the casualties at sea; from which it appears that, among 11,468 men, only thirty-three deaths occurred, six of whom were accidentally drowned; nine of whom died consumptive; two from hæmoptysis; two from aneurism; and in fact none from disease originating on board ship.

With such facts before us, it is well to pause before recommending so great an additional public expense as the entertainment of a fleet of transports, for as one of the authors just quoted (Dr. Chevers) truly observes, "these facts tend to show, not only that the arrangements of Government for sending out our troops have long been as nearly perfect as it is possible to render them, but also that the English soldier is healthier when on his way to India on a well-formed ship, than in any other known position in which he can be placed."

However ships are procured, it is admitted that on long sea voyages, vessels ranging from two thousand tons register and upwards, furnished both with sails and steam-power, are best adapted for the transport of troops. The decks should be kept free from impediment, and the vessel furnished with ports for ventilation. Store-rooms, hospital, dispensary and surgery, school-room and library, galleys and cook-house, laundry and wash-house, mess-rooms and dormitories, should be fitted up in the interior. Care should be taken that the decks are properly caulked and not admitting leakage; life-boats, with patent lowering apparatus, should be attached; force-pumps, for clearing side-houses and urinals, supplied; risk of fire efficiently guarded against; the sufficiency of night lights ascertained; shoots for discharging refuse inserted at a proper angle in the ship's bulwarks, and a due supply of canvass awning be taken on board, made from new cloth, and not from old and partially worn sails.

As Dr. Kirwan observes, no vessel which has been built for the purpose of trading in northern climes is adapted for con-veying troops through the heat of the tropics, and from their original construction it is almost impossible to alter them so as to render them suitable for sea-barracks for soldiers during a long voyage.

It is generally admitted that as few articles as possible of live stock should be taken on board. The author just quoted remarks, "such an animal as a pig should not be allowed on board a troop ship. The amount of dirt which they create, and the intolerable stench which is ever about them, forbid their presence."

It would, however, take up many pages of this book to enter into a full description of all that is required for a troop ship. It is sufficient to say that every ship should be of the first class, and should have stood the test of former voyages;

for, as Blane long since observed, " infectious ship fever is the disease of vessels newly fitted out and built of unseasoned timber;" and " that every appliance for the health and comfort of the men with which modern science has made us acquainted, so long as these do not degenerate from comfort into luxury," should be provided.

During a voyage with troops the medical officer should be the sanitary officer, and on him will devolve the duty of seeing that the bedding is frequently aired ; of recommending the removal of nuisances and matters deleterious to health ; of regulating the spreading of awnings, and the number of men to keep below or on deck ; the superintendence of ventilation ; the inspection of the men for symptoms of scurvy, and the insisting upon its appropriate remedies being issued ; the examination for skin diseases and other contagious affections ; the segregation of the sick ; the insurance of cleanliness and exercise; the supervision of the diet ; and also the administration of the appropriate remedies for individual cases of disease.

Mr. Kirwan is of opinion that night watches on deck are a fertile source of sickness in troop ships; but I cannot say I have ever found this to be case when the men sleep or watch on deck with awnings above; in fact, I believe with Major Haley,[1] after some experience of this matter, that much disease is engendered by sleeping below in warm weather in a crowded and over-heated atmosphere, and that sleeping on deck (under an awning) should not only be encouraged but *enforced* during the warm weather and calms, to the extent of at least two-thirds of the number.

Any medical officer who has served in the Persian Gulf, or Red Sea, as I have done, will admit the impossibility of a ship's crew sleeping below, although, indeed, they might *be there by command* during the night, and suffer from disease,

---

[1] ' Suggestions for the Dress, Camp Management, &c., of the Army.'

as was the case in one of Her Majesty's ships of war, which some years since cruised in the waters of Iran.

During my period of service in the Indian navy, whether in charge of troops, or simply with the ship's crew, I invariably recommended that the awnings be spread at night during the hot months for the men to sleep under. Nautical men, however, appear to have an insufferable objection to spreading awnings at night—the reasons advanced for non-compliance with such a recommendation from the medical officer being, that a sudden squall might gather under them; that a spark might set fire to them; that it is not man-of-war like; and that exposure to the dews of night rotted the canvass.

During some portions of the year in the tropical seas, the frightful heat of day is succeeded by copious deposit of dew during the night. Any one sleeping under the star-lit canopy of heaven only becomes deluged with wet; if he " wrings " his garments water falls from them as though just removed from a well; and handfuls of fluid may be scooped from the ship's scuppers. What wonder then that the man rises (as I have frequently done) cramped, cold, with aching pains in his limbs, and disgust in his heart ! What wonder that the seeds of rheumatism, of albuminuria, and other chronic diseases, enter into the system, destroying health almost before manhood is arrived at !

The objections to spreading awnings mentioned above are not worthy of one moment's consideration, being simply absurd and obstructive. Another objection which may be advanced with more force is, that awnings prevent the thorough perflation and ventilation of the ship. Therefore one hour before sunset in the tropics, the awnings covering the deck by day should be furled and remain so furled, until one hour or one hour and a half after sunset. This would afford ample time for the breeze which generally springs up in the evening to ventilate and cool the ship. The same process should be repeated before and after sunrise in the morning.

Although this furling and spreading awnings should always form a portion of the daily duty of the ship, still it is not actually and positively necessary for the sake of ventilation, which may be thoroughly effected by a sufficient number of adequate and properly shaped windsails. These canvass chimneys should penetrate through the hatchways, or through openings in the deck made on purpose for them, to the extremest depths of the ship, and they should pass *through* the awnings above in order that their mouth may be fully exposed to the prevailing breeze. The plan of allowing the mouths of the windsails to open underneath the awnings renders them of little use, but I have frequently seen this done in that now moribund service the Indian navy, on the ground that it was not man-of-war like for the windsails to be above the awnings. Such frivolous objections, such sacrifice of health and comfort to nautical " smartness," cannot be permitted on board a troop ship, neither should the caprice of an " executive officer " be allowed to render the life of a whole ship's crew miserable, and their bodies predisposed to disease.

When on board steamers a head wind drives smuts and blacks from the funnel into the wind sails, the mouths of the latter must be carried away from the centre line of the vessel, sufficiently on one " quarter" to avoid this unpleasant result. Hence an extension of the windsail will be required, which should be constructed to fix on to the other part.

Another fertile source of disease is the washing and sousing with water which most naval men appear to think essential to cleanliness. In the cold weather of the Persian Gulf, the Red Sea, and other tropical climates, ships are kept constantly damp from this cause. No sooner has the meridian sun dried the decks from the morning deluge, than the evening dew again saturates them, the completion of the process being effected at the dawn by bucketsful of water handed over the ship's side. So long as dry scrubbing will suffice, the main and lower deck of a troop ship ought not to be deluged with water, and it would conduce to the health of ship's companies generally if a little less attention were paid to the sweeping and whitening of the upper decks, *and the surplus labour expended on the hold and other dark, dirty, and interior spots.*

From recent occurrences, as that of the Tasmania, that of the Clyde, with the 64th Regiment on board, who on account of the bad sailing qualities of the ship, were so long in reaching England from Kurracchee, and other instances of a similar nature, it would appear that troops arriving in England generally do so after a more tedious passage in many respects than is the case on the outward voyage. The cause of this is sufficiently evident. The colonies do not possess the same resources either in ships or supplies, and hence without Government transports it is no wonder that from the large number of vessels chartered for the convoy of troops, we occasionally hear of slow sailers, impure provisions, and their consequences. Here is a more cogent reason for the establishment of a fleet of Government transports than is to be found in the mortality of soldiers on board ship during their outward voyage.

Although it has been demonstrated that sickness and mortality amongst troops at sea is less than in most other positions in which soldiers are placed, still the same favourable results are not found to occur when children are the subjects. The mortality of the latter on board ship, probably from the inconveniences they there undergo, and from the food not being adapted to the earlier years of life, is frequently excessive. (See Chapter XXVIII, "On the Soldier's Wife, Widow, and Children.")

CHAPTER XXVII.

ON EUROPEAN COLONISATION IN INDIA.

THE subject of colonization, and especially military colonisation in India, has been frequently discussed. It was the opinion of some of the ablest writers on India, including the late Sir H. Lawrence, that it would be both politic and practicable to establish colonies on the slopes of the Himalayas, where soldiers on retiring might be induced to settle down in peaceful pursuits. The policy of this arrangement is unquestionable. It would cause a large class of Europeans to take root in the country, and encourage the growth of useful institutions. It would prove a source of strength in a possible hour of need; and who, after the experience of 1857, shall say when that hour may not be at hand?

The practicability of the scheme, however, is not so certain, and as regards carrying it out by means of military colonisation, must appear to all who understand the European soldiers of the present day most impracticable.

A British soldier who has been accustomed to be looked after all his life is a helpless being when he has to look after himself. He is as much out of his element as a sailor fifty years since used to be on shore. He is, to use the words of a late writer, like "an old charger, who wants the trumpet and the word of command, and does not understand kicking up his heels by himself; with the rest of the troop he will go through his work with credit, but once alone he falls into bad paces, and discovers weaknesses of which he never dreamed."

Moreover, take the European soldier who has served his period of service, obtained his pension, and therefore is eligible for a military colonist; let his physical condition as regards health and disease be reported on, in short, let medicine give her opinion as to the qualifications of the man, for the rough work of a colonist.

The answer would probably be, he is old; he is debilitated from tropical residence, heat, and malaria; he is unfitted for hard labour by these causes, and because he has not throughout his military service been habituated to dig and plough; he is subject to intermittent fever; he has an enlarged spleen; he has suffered from liver (quite a sufficient reason for his avoidance of a hill climate); he has had dysentery, from which he occasionally now suffers—in fact, he is totally unfitted for the life of a settler or pioneer of civilization.

It is not intended to assert that *every* soldier who becomes eligible as a military colonist would be unable to adapt himself to a new mode of life, or to bear the fatigue of a labourer, or that even *all* would be diseased. Now and then exceptions might occur, but such would scarcely exceed five per cent. By far the greater number would find a labouring life irksome; would perhaps fall victims to their own evil passions, of which intemperance is the most prominent; and a still greater proportion would be bodily unfitted for the new mode of life, by physical degeneration from climate.

The first objections might be certainly overcome by placing the colony under military regulations, sowing and reaping by word of command; but such an institution would not be popular with the soldier, who looks forward to perfect liberty as the termination in his pensioned career. The second objection—the physical degeneration or disease—cannot be overcome even by military discipline.

Hence it may be asserted that practically all our soldiers, who, by length of service, are qualified for pension, are by the same period of servitude, by disease, or by degeneration from residence in the tropics, unfitted for the life of a colonist, either on mountain ranges or elsewhere.

In addition to the advocacy of colonisation on mountain slopes, it was formerly thought that the same might be attempted on the plains, and the healthy appearance of planters, &c., has been brought forward in proof. It must, however, be recollected that such gentlemen are not daily labourers, that they live in magnificent houses, with every comfort surrounding them, and that although considerably exposed to the sun, their duty is that of inspection, and not the rough manual labour of the colonist.

In the absence of any statistical or authentic data, showing the state of health of indigo and other planters; knowing that the chubby or ruddy appearance frequently seen in India, caused by the expansive powers of heat and flow of blood to the cutaneous vessels, instead of being indicative of health, is commonly the reverse; knowing that heat and malaria will induce their results in the human constitution, *cæteris paribus*, whether the individual be soldier, sailor, or civilian; being aware also that it was "at home" after months or perhaps years of absence from the tropics, that the healthy appearance of planters was remarked upon:— bearing all these facts in mind, I must decline to believe that one class of Europeans in India is less mortal than another, or less liable to the endemic diseases of the climate.

It is the fashion now in some quarters to declare that the dangers of Indian residence and service have been greatly overrated, and that there is little or nothing of an exceptional character in the climate of India to render it necessary that special inducements should be held out to persuade Europeans to reside therein, and this too in the face of so many stern facts which have confronted us within the last few years. How many more victims must be added to the list of those killed by climate before the dangers of tropical residence become appreciated?

If colonisation, as America and Australia have been colonised, were possible in this country, some instances of the kind would already have occurred. But the melancholy truth is, that the European race dies .out. Of the numerous pensioners who have settled at our principal military stations, how many have been colonists? There is not one single instance! There is not a great-grandchild or grandchild of these pensioners retaining their European characteristics! An infusion of native blood is essential to the continuance of the species, and the barrier once broken down, the remoter descendants of an European ancestor become rapidly feeble, astute, passionate, and indolent, as any of the darker races around them.

The fact is, for the white man or his offspring there is no such thing as acclimatization in India. As a rule, Europeans enjoy the best health, and suffer less from heat during their first years of residence in this country. Acclimatization as regards an Indian sun is simply impossible. Exposure, instead of "hardening" the system, actually has the contrary effect, and the *longer Europeans remain in this country the more they feel the effects of the vertical sun.* Men with a larger amount of strength and vital force than others can bear exposure and the effects of heat longer than those not so gifted; but the deteriorating process, though slow, is nevertheless certain, and if acute dysentery, epidemic cholera, ardent fever, or sunstroke do not some day suddenly destroy, insidious malarious disease, cachexia loci, or splenic leucocythæmia sooner or later results.

When Europeans urge that they have exposed themselves

to the sun for years, and have never felt any evil effects, it is only saying that the losing battle between the sun and their constitution is not yet over, but every day's exposure brings them nearer to the final triumph of their solar adversary. The lamented fate of that gallant sun-defier Colonel Jacob, who advised young officers not to mind the sun, as it "would only tan their cheeks," is an apropos example of the foregoing.

A tropical climate, in addition to the depressing and deteriorating effects of heat and malaria on the system of settlers or dwellers in such localities, presents another indirect action in aiding the same result. In hot countries the animal heat is more easily kept up; and as on the whole the bodily exertions of the inhabitants are less frequent, and the decay of tissues therefore less rapid from the latter cause, a smaller amount of carbonized food is required to keep up their heat, and less also of azotized food, because there is less habit of bodily exertion to cause waste. In cold countries, on the other hand, men are compelled to eat more than in hot ones; while their food from its necessary quality is more expensive and less easy to obtain, and requires a greater expenditure of labour. This degree of exertion in a rigorous climate imposes the necessity of a diet of a more abundantly carbonized and nutritious quality in order to sustain heat, and compensate waste and attrition. In warm climates, as Mr. Buckle[1] observes, sufficient food is supplied to the inhabitants " by the bounty of nature gratuitously and without a struggle," less labour is necessitated to procure the necessaries of life, the vital forces are depressed by heat, and hence the apathetic condition of the natives of tropical climates.

Mr. Buckle selects Hindostan as an example of the foregoing, and adverts to the great heat of the climate as bringing into action that law by virtue of which the ordinary food is of an oxygenous rather than of a carbonaceous character; thus compelling the adoption of a diet rather from the vege-

---

[1] Buckle's ' History of Civilisation.'

table than from the animal world. The chief food of millions in India is rice—a grain which, if not one of the most nutritious, is certainly perhaps the most easily produced of all the cerealia. Hence a rapid growth of population, a redundance of labour, and that consequent inequality in the distribution of wealth which tends to render the upper classes enormously rich and the lower classes miserably poor, fitting the latter to be little better than slaves. This condition in India Mr. Buckle designates as an abject, eternal slavery—a state to which the inhabitants are doomed by physical laws utterly impossible for them to resist.

There is not an instance on record, Mr. Buckle continues, of any tropical country, in which wealth having been extensively accumulated, the people have escaped this fate; no instance in which the heat of the climate has not caused an abundance of food, and the abundance of food an unequal distribution, first of wealth and then of political and social power, terminating in degeneration of all classes.

As it has been with inhabitants of tropical nations, so will it again be. In addition to the effects of malaria and heat on the constitution, this indirect result of climate forbids the hope of colonisation of the Indian plains by Europeans.

The effects of heat have been several times referred to in this work, and a few remarks on the subject may with propriety be introduced in this place. Heat causes a relaxation of the tissues, and produces at first excitement of the nerves and circulation, then languor and weakness of the whole system. The mechanical or direct effect of heat, as previously referred to, is to expand and thus diminish the healthy tension of the tissues, while, at the same time, it induces a flow of blood from the interior towards the surface. This excitement, however, is soon succeeded by general weakness and lassitude, and increase of perspiration. The natural heat of the body is 98°, so that when the air outside the body is warmest, the latter exerts its means of producing cold, and this is afforded by the evaporation of the perspiration exuded. Excessive perspiration, however, always tends to exhaustion, so, therefore, in this is an additional cause of languor from external heat.

Another result of heat is an increase of the waste of the body. Heat promotes the transmutation of tissue in the living animal as in the dead, and augments the proportion of those excreted and effete matters, either in the urine, the perspiration, or the alvine dejections, which are constantly passing from the muscles, nerves, and other portions of the frame.

Exercise, as noticed in Chapter XXIII, " On Dress," increases the effects of heat in a tenfold manner. Colonists cannot live without exercise, without, indeed, hard labour. The constitution, the sensibility of the skin, and the former habits of the European, forbid such labour, even if the additional cause of malaria did not destroy his existence at an early period.

Not a single reliable fact has been produced to show that our race can be continued even through a few generations without Asiatic mixture. The children of Europeans born in the plains of India grow up weakly, the progeny seldom attain maturity, the third generation never have children.

All authorities agree in stating that not one descendant of the Portuguese can be found without admixture of native blood.

Henry Marshall shows even the domestic animals of colder climates, dogs, cows, sheep, horses, all die or deteriorate after removal to a tropical region. Hence, colonisation of the plains either by military pensioners or otherwise is impossible.

The question, however, is not yet decided if a *healthy and vigorous* European stock can be propagated and maintained in the mountain climates of India. Sir R. Martin, in his answer to one of the questions of the select committee on

Indian colonisation, states that this might be the case " to a certain degree not yet determined." Bakie, before the same committee, had "no doubt a race of persons well off in life would be continued on the hills," but whether they would deteriorate was a question he was not prepared to answer— "nothing but time can solve that question."

Dr. Mackay,[1] in his paper on the climate of the Neilgherry hills, reports, " children brought up here apparently strong and healthy in their youth, in after years show constitutional weaknesses; and this I have observed to be the case particularly with females."

The same is the fact to a certain extent with the children in the Aboo Laurence school, and I believe in all mountain asylums.

It must, however, be recollected that these children are *born* on the plains, and remaining there some years before they are admitted into the hill schools (limit five years of age), generally arrive at those establishments more or less deteriorated by former residence in malarious localities.

Were they the progeny of healthy parents, born on the mountain range, into a locality from which sanitation had swept away the present sources of disease, there is every reason to believe that the vigour of the race occasionally renewed, *as is the case in America* by new blood, would not degenerate, were the studies of such colonists confined to superintendence. Whether, however, European progeny would retain their characteristics if obliged to undergo the exposure and labour consequent on tilling the ground of even hill ranges, can only be determined by time and experiment. With constant infusion of new blood, and due attention to sanitary principles, there is ground for the belief that such might be the result.

However this may be, one thing is quite certain, that the hill climates of India have never been yet sufficiently tested,

---

[1] 'Mad. Med. Journal,' vol. v.

have never had their capabilities thoroughly developed by all
that sanitary science is able to accomplish. This is sufficiently demonstrated in the chapter "On Hill Sanitaria."
Formerly it was a favourite allegation that, however excessive might be the mortality of any district, no evidence was
forthcoming to prove it due to local preventible causes, and
that it was in every case not the accident of the employment
of the population, the peculiar position of the town, and
other matters, which were beyond control and irremediable.
In no one thing have sanitary reformers accomplished greater
results than in proving that mortality ever and everywhere
bears an inverse ratio to the amount of sanitary supervision
and care in any given district; those accustomed to the
statistics of disease, its causes and prevention, need but an
inspection of a locality to form a very close approximation to
its absolute and relative immunity from disease and death.

I repeat the assertion made in the chapter " On Hill Sanitaria," *that sanitation in its broadest sense is more required in
the climate of tropical mountain ranges than in any other
known locality.* With due attention to such matters there is
reason to believe that colonisation may be effected on Indian
elevations; any colony or community existing without such
care, *would most assuredly eventually become the victims of
the most deadly diseases of both temperate and tropical climes.*

It is, however, questionable if suitable or remunerative
employment could be found for a large number of settlers
on the mountain ranges. For many, certainly, such employment might be found in the cultivation of tea, coffee, and
cinchona; but much of the hill side is rock, or too steep to be
susceptible of culture, and many of the valleys—from causes
previously referred to—are so deadly malarious, that the
pioneers of cultivation and civilization would be destroyed
before their efforts succeeded in transforming the valley, by
drainage and clearing, into a healthy locality.

There is more difficulty in bringing a hill side under tillage than a piece of level land; such cultivation can only be effected by means of terrace work, and the latter requires much care in its construction, and is liable to be washed away during the violent periodical rains. Thus the settler in a mountain range *would require capital*, and such capital will afford greater profits employed in procuring the labour of the natives than that of Europeans, the former, although, not capable of performing so much actual work individually, from their number and fewer wants, defying European competition.

It is therefore evident that the fostering care of Government must be extended in the initiative to hill colonies in the broadest sense of the term, by which some of the following results may be achieved :—

The breeding of sheep in the Himalayas, where the pasture grasses are unsurpassed, and the sheep or purple *fescue* indigenous, and the climate presents that dry cold so necessary for the perfection of wool.

The breeding of the shawl goat of Cashmere in the same localities.

The cultivation of tea, coffee, cinchona, oranges, vines, and olives, in the Himalayas, the Neilgherry, the Shevaroys, the Avarallee, and other ranges.

Some of these localities present apparently identical climates with those of the Cordilleras of the Andes, the locality in which Peruvian bark is indigenous.

The great importance of introducing the cinchona plant into India, has become more and more apparent of late years owing to the increased demand for its invaluable alkaloid quinine. Dr. Ewart [1] shows that the expense of this agent alone during 1857-8, for the Calcutta Presidency, amounts to £12,408, besides the cost of thousands of pounds of cinchona bark, and upwards of one hundred pints of liquor amorphus

[1] " A Review of the Treatment of Tropical Diseases ;" ' Ind. An. Med. Sci.,' vol. xiv.

quinæ. In Madras, during the same period, 267 lbs. of quinine was used, costing £1,284.

The Dutch have already succeeded in growing the cinchona plant in their eastern possessions, and the introduction into India was commenced under the fostering care of Lord Stanley, when Secretary of State. Recent accounts show that there is every probability of the cinchona plant being naturalized on our Indian ranges—an event which will be trebly beneficial, relieving the state of considerable expense, affording a wide field of occupation for European colonists, and rendering the antiperiodic cheaper and more available to all classes resident in India.

CHAPTER XXVIII.

ON THE CONDITION OF THE ANGLO-INDIAN SOLDIER'S WIFE, WIDOW, AND CHILDREN.

Mortality and Suffering from Climate—The Unmarried Soldiers—Means—Can the European Woman increase her Income?—Proposed Allowance—Indian Mutiny Fund—The Anglo-Indian Army Wife and Widow Fund—The Soldier's Wife's Habitation—Married Quarters at Dugshai—The Soldier's Widow—The Soldier's Child—His conveyance to Europe.

It has already been shown (Chap. V) that the mortality of European women in India is in Bengal 44·4 per 1·000, and that of children in the same Presidency 81·2 ; the mean of the three Presidencies being for women 35·477, and for children, 64·322 per 1·000.

With such a death rate the suffering from climate and sickness must be beyond imagination !

There is no one single matter connected with our Anglo-Indian army more demanding attention than the condition of our soldiers' wives and children, and this notwithstanding the ameliorations which have been latterly effected in their state both by private benevolence, as that of Sir Henry Laurence, and by the humanity of Government, as instanced in the grants to Laurence Asylums, and by the different orders relating to the subject.[1]

The condition of the European woman must be considered under two different phases of her existence—namely, that of the wife and that of the widow.

An unmarried soldier can just exist on his pay. If he

[1] *Vide* ' Bengal Mil. Reg.,' sec. xi, p. 64.

has a wife the State allows her rupees five per mensem if she be an European, and two and a half rupees a month if she be of country birth. This is of course conditional on the assent of the commanding officer having been obtained prior to the marriage. Each child is allowed half the subsistence money of its mother. On this, with the soldier's pay (18 rupees per month, without rations), the family manage to exist— how is another question. That life is *not long* is sufficiently evident from the death rate referred to.

But it may be expected the woman earns something by her own exertions, and this she can perhaps do, if she has no children, to the extent of the wages of a London shirt-maker!

Opportunities of increasing their income do not present here as they do at home to women, and if such were available, the climate forbids their practical application. As instances in the most common feminine occupations, the European woman cannot compete with the Indian " dhobee," or washer-man ; with the " durgee," or tailor; with the " bowarchee," or cook. She cannot become a charwoman, butterwoman, huckster, or labourer—in fact, there is no occupation for her in India.

There is no European regiment in India where money is not paid to the native " durgee" for needlework for the officers' family, to the washerman, to the native ayah or nurse. Could these wages be diverted to the European women, how much improvement would it not effect in their condition? There are, however, several cogent reasons which oppose the general practicability of this. A woman with a family has no time for needle or other work—the woman without a family can live better than the former, and in the majority of instances, from the effects of climate, becomes apathetic, or if induced to work, asks, as her husband has been stated to do (Chapter XXIV, " On the Employ-ment of the Soldier"), enormous wages for her labour.

Unless soldiers' wives are first allowed the means to obtain that position to which self-respect sufficient attaches to incite them to efforts to retain it, the great majority will never be otherwise than they now are. It cannot be questioned that the allowance to soldiers' wives and children is meagre in the extreme—in fact, is not enough for their *wants* in a tropical climate, and under the circumstances related, how is she to be expected to better her condition by individual exertion?

Fifteen rupees a month at least ought to be allowed to the European woman in India, and opportunities *found* for her to earn as much more. Thirty rupees is none too much for an European female to live upon and clothe herself with in this expensive country.

Why should not the remainder of the Indian Mutiny fund be set apart for the purpose of adding to the Government allowance? nay, why should not a special fund be raised for this purpose? The " Anglo-Indian Army Wife and Widow Fund" would be as deserving of support as any missionary society or benevolent institution in the " wide wide world."

Next to the provision of daily sustenance the question of habitation for the European woman arises. It is to be hoped that in no station is it as Lord Carlisle wrote but a few years since, " At present they for the most part sleep in barrack rooms with the rest of the men ;" or that, as Mr. Godwin observed in 1859, " two or more families are lodged in one room, with no other separation than the curtains drawn round the bed."[1]

Dr. Chevers[2] informs us that for some years past accommodation for married families, entirely apart from the unmarried soldiers, has been provided in nearly the whole of the Bengal military stations.

In many stations, however, the European women inhabit the buildings known as " Patcherries," which Dr. Arnott

---

[1] Evidence before the late Lord Herbert's Commission.
[2] 'Ind. An. Med. Sci.,' vol. xii, p. 714.

describes as " too often most mean, uncomfortable, and unseemly appendages."

Patcherries are constructions of different forms, sizes, and dimensions, and appear to have derived this name from the situation which they occupy with regard to the barracks. " Peecharee" in the Hindoostanee language signifies the rear or behind, and " Patcherry" seems to be a corruption of the former. The habitations of soldiers' families being generally erected in the rear have hence come to be designated by the latter term, even when their relative position may be in any other direction, and whether such dwellings may be under one roof or detached.

Mr. Bradshaw, writing to me from Dughshai, thus describes the married soldiers' quarters at that place :—

" There are two barracks for married men, each intended for thirty families. Their positions are very good, but the plan on which they are built is remarkable for its want or deficiency of common sense. Each barrack is the ordinary ugly barn, surrounded with a verandah, and divided length-wise by a long tunnel-like passage, which communicates with the verandah at each end, and on each side by two narrow passages. In the tunnel are six fireplaces, and into it open the doors of all the rooms. These rooms are small, incom-pletely divided by makeshift partitions, and insufficiently lighted. The women use the fireplaces for cooking and other purposes, and ordinarily the passage is quite dark from smoke and unpleasant from culinary odours, although each of the barracks has a roomy cooking house belonging to it, and also a detached rear, &c.

" None of the rooms communicate directly with the verandah, and this is closed at each end by rails. Imagine thirty families in a house with only four narrow passages as outlets. . . . The houses can hardly be otherwise than unhealthy, or deserving of any other name than kennels, or other fate than being pulled down."

Without multiplying descriptions of this nature it will be readily observed, that the difficult problem of lodging the married people of a regiment serving in a tropical climate in such a manner as to secure at once the maximum of health, space, privacy, supervision, cleanliness, conservancy, ventilation, and coolness, has not yet been satisfactorily solved.

As with barracks there is only one plan which presents the maximum of these desideratives, and that construction consists in the extension of one line of married quarters to the prevailing breeze, and by *avoiding all double ranges and cross partitions*, or *otherwise*, by providing upper sleeping apartments.

The standing ground plan of a barrack for thirty married men as prepared in the office of the Public Works Department in 1857, giving a length of 433 feet, including verandah, may be taken as the nucleus. This should be raised on arches, contain one room, and a small enclosure, to be used as a pantry, &c., in the rear corner, with doors and windows opposite each other. Above this should be sleeping apartments, the size of the lower rooms, but not including the verandahs. A tatty for one door or window of each apartment, all of which might be worked by steam or other mechanical power, would be sufficient to cool the lower story during the hot weather. The partitions between each apartment should consist of two single brick walls, with an interval equal to the breadth of one brick *between each wall*, which space should communicate by several apertures with the external atmosphere.

At the distance of ten or twelve yards to the rear of each house a necessary and washhouse should be built; and a covered passage, made simply of uprights and matting for the roof, without sides, should form a protection from the sun when proceeding to and from the dwellings. These outhouses should be cleansed daily by the public conservancy establishment, and the whole kept sweet and pure by a

19

thorough system of cleanliness and the use of disinfecting agents. To the nature of the ground allowed, drains for the exit of water should be excavated, otherwise manual labour must be had recourse to for the purpose of preventing sodden ground, and its sequel, malaria.

Of course such a range of buildings, with appliances and establishments, would be somewhat expensive, *but, as is the case with the unmarried soldiers' barracks, the choice only lies between the Scylla of expense, or the less desirable Charybdis of disease and mortality.*

The condition of the soldier's widow may now be considered. If the husband die or be slain, his widow is allowed her stipend for six months. " Out of this thirty rupees she may if she has any feeling erect a tombstone, and must pay off her little debts, find and furnish herself a home, and console herself as best she may." When she has received from Lord Clive's Fund that sum of money, she has by law no further claim in respect of what was *her all.* The dead man's dog is turned out of barracks if no one else will accept its ownership, and so is his wife! If she has no friends she must starve, sin, or marry, and rid the community of an incumbrance at once. Means of earning an honest and comfortable single living there are in India none for her; when she had the poor merit of owning a living husband, she could scarcely find the means of earning; what can she expect as a widow? Two means of life only are left her—to do violence to a woman's feeling in one way or other, or to starve !

Supposing she has children her case is fourfold worse; their pittance would scarcely support a village "ryot." A widow with a daughter fifteen years of age has seven and a half rupees to live on. Both have the alternative of marriage or starvation at six months' distance before them. They must either marry or do worse !

Here is call for a soldier's widow and orphan fund. Too much time has been already lost. Let those with power and means delay no longer! Let those who are foremost to pride themselves on English civilization, honour, and gallantry reflect how strongly those qualities are exemplified in the treatment of the British soldier's family. Let all who strive "to do unto one another as you would they should do unto you," think and act. Let pity and piety be more practical hitherward than heretofore. Let the British nation, the merchant princes of England, those whose wealth is secured by the British army, establish "The Anglo-Indian Soldiers' Wife and Widow Fund." Never was a time when the wants and comforts of man are more largely administered to than during the present age. Look at the efforts for the relief of human suffering—when were they ever so great as now? Let our hospitals, asylums, infirmaries, and charities, our missionary societies—*nay, our very prisons*—let them all bear witness to the interest now taken by man in the welfare of his fellow-men, and to the practically humane tendencies of the age. *The wealthy British public may well be called upon to add one more institution to the number—an institution which is as much required as many now upheld by humanity and charity.*

In a strictly military sense women and children attached to a regiment cannot be regarded as anything but incumbrances; but we must accept the difficulties of our position and endeavour to meet them as best we may. A regiment is allowed so many buglers, so many officers, so much baggage, so many beasts of burden, and so many wives—the latter amounting to 12 per company. The "holy estate which He himself adorned and beautified by His presence is proscribed," except under circumstances of much difficulty to the man who humbly serves his country. As, however, I stated in my ' Manual of the Diseases of India,' page 33, instead of limit ing the number of married men to so few per company, I would allow without restriction all men of good character to contract matrimony when opportunity offered. This would be one of the most certain methods of preventing ennui, and its sequels, intemperance and syphilis.

I am aware of the objections which may be brought forward against increasing the number of European women beyond the regulated proportion, the principal being the increased expense they would necessitate, the want of accommodation for them, the inconvenience they are subjected to when the regiment is frequently moved, the mortality among them induced by climate, and the prevalent idea that a married soldier is less efficient than one without incumbrance.

Regarding the first objection, viz., increase of expense, I can only observe that supposing one man only in any moderate given number was reclaimed by marriage from drunkenness, syphilis, and death, the expenditure thus saved to the state would suffice to cover the additional expense of many married women.

The want of accommodation is certainly a graver objection, but not irremoveable, as the barracks which are being erected every day might be shorn of their proportions, and the materials employed for the construction of married quarters.

Of course when a regiment moves the women and children must be subjected to inconvenience, but in no station of life would their path be without thorns. Moreover, with returning peace, military necessities will not be so urgent as heretofore, and if in the course of years state affairs allow of most regiments having their home or depôt in one or other of the hill stations, the families would remain there if the regiment was called upon to march at unhealthy periods.

The last objection is entitled to little weight. As a rule the well conducted men are the married soldiers, and these knowing their wives and families were in safety, would not shrink from their duty in the day of trial. By these means, by increased domestic comforts, and by residence in hill stations, the mortality now so great among European women and children would be reduced to such an extent that their condition *would be infinitely preferable* to that of their friends at home in the large manufacturing and seaport towns of Great Britain.

With regard to children it has been proposed that they should not be brought from England when a regiment leaves for this country; and as in the majority of instances they only arrive in a tropical climate to die, humanity forbids that these little ones should be any longer disposed of in such a manner. Statistics and the general experience of medical officers combine to show that children born in England die when placed in barracks in India, and that whatever be the mortality of infants born in this country, *it is considerably less than that of those brought here.*

Dr. Chevers, after detailing statistics bearing on this subject, remarks:—"These facts alone certainly justify the question whether the humane intention of reuniting the soldier in India with the little children for whose presence his heart has so often yearned, is not at the very onset more than forbidden and nullified by the fact that one quarter of these tender bodies may probably be committed to the great deep during the four months' voyage."

Mr. Walker of Bombay having become aware of the great mortality of European children in India, has latterly brought forward a plan [1] for sending children of European parents to England to grow up and be educated there, and states that if the revenue of the Byculla, Poonah, and other schools were diverted to this use, £22 11s. 7d. per head per annum would be available.

Mr. Walker observes : " Who will deny that with such a revenue as this the children would not be well fed, clothed, and educated in a fine healthy part of Yorkshire, where food, fuel, and clothing are cheaper than in any other part of the world ?

[1] 'The Times of India,' Nov. 25, 1861.

" I have no doubt the average cost of all children in India is more than £20 a year (Byculla school children cost last year £22 per head); the infants at Poonah cost £20 6s. each per head, which is a high rate. I would engage to place them out in country places with good motherly dames, having children of their own, for less money, where they would be as rosy as pippins and merry as crickets;—*we never see a child making a mud pie in India.*

" It will, with these facts before us, be an easy question to solve, by reference home to some of our chief orphan schools, to ascertain whether that sum is not ample for the purpose when conducted on such a large scale.

" The gain to the commonwealth in having strong and healthy, instead of puny, debilitated children growing up to serve the State, would be worth ten times the petty extra cost, and we shall have the satisfaction of knowing that we have done our duty to the helpless orphans."

As the editor of the 'Times'[1] of India observes:—"The proposal which Mr. Walker has started with regard to the children of the lower class of Europeans born in this country is one of really national importance. It is all very well for people to look at the children in the Byculla schools or any similar institution, and to say they are very happy and do very well where they are. Take those children in ten years time, and then decide whether they are equal, either physically or mentally, to what they might have been, had they been sent home for their adolescent years. Take their offspring again, if they have any, and judge whether a natural tendency, in fact a natural law of degeneration, has not plainly manifested itself?

No one disputes the desirability, the necessity, in fact, of sending children of British blood to England. Every one who can by any means compass that object does so for his own progeny. As Mr. Walker observes:—"To a rich nation

[1] 'Times of India,' Dec. 23, 1861.

like ours it is certainly a disgrace to let the bone of our bone dwindle away till they become, like the descendants of the Portuguese at present in India," a bye word and reproach of the heathen around.

There are only three courses open to us as regards the disposal of European children in India:—either they must be allowed to die at an early age as heretofore, or to live and become the degenerate race every one laments in the present Portuguese of India ; secondly, they must be located on hill stations at an early stage or from their birth; or thirdly, they must be sent home before they are deteriorated by climate.

The first proposition is against all dictates of humanity and charity, and will render humane and wealthy England centuries to come the compeer of impoverished and bigoted Portugal.

The second and third propositions should both be adopted; those children who cannot be, or who, from medical reasons, are unfit to be, located on hill ranges, should be sent to Europe. England is sufficiently rich to carry out both these essentials, and their advocacy is only required by some one great and powerful, and to such I would write—

> "Mark, when the life of man is in debate,
> No time can be too long, no care too great !"

A superficial observer may, however, exclaim, the mortality on board ship consequent on bringing children to India has been advanced as a reason against such transport, and yet at the same time their conveyance to England by the same means is advocated. Such objection is, however, disposed of, *paucis verbis*. The mortality of children coming *from* a temperate climate to the tropics would be in an inverse ratio to those proceeding *from* the latter latitudes *to* the former.

CHAPTER XXIX.

REMARKS ON QUARANTINE.

THE history of quarantine does not date from any remote period. It was about the year 1448 that the first code of quarantine regulations was promulgated, and that in the same city in which, a few years before, the first lazaretto was established. This city was Venice, then the great emporium of eastern trade, and hence most in danger of suffering from contagious diseases. Since that period quarantine regulations, more or less strict, have prevailed in all countries situated within a certain latitude, and having any claims to civilization. Of late years, however, the subject has been much agitated, and many able men, both of the medical profession and otherwise, have advocated the total abolition of quarantine, on the grounds that such regulations have not worked beneficially, have not prevented the spread of epidemic disease, have proved most injurious to commerce, and been the source of vexation, annoyance, and even disease, to enormous numbers of travellers.

Regarding the first objection, it is very certain that cholera, plague, and other kindred diseases, spread otherwise than by contagion; but on the other hand, it is again equally well authenticated, that individual cases of such maladies may become the *fons et origo* from which the disease spreads by direct or indirect contact in every direction; and this especially when the locality in which a disease is introduced is wanting in all sanitary characteristics, as is so frequently the case in eastern cities.

Dr. Bryson, in a valuable paper read at the Epidemiological Society Meeting, March, 1861, supplies a valuable chain of facts relating to the successive outbursts of yellow fever in a consecutive series of Her Majesty's ships. The Icarus had touched at Belize in August, where the yellow fever raged. Soon after leaving, the crew were attacked with the disease, and out of 102 persons thirty-seven died. When the vessel reached Port Royal, a boat was sent from the Imaun, to assist in landing the crew. Eleven out of fourteen of this boat's crew were seized with the disease, and most of them died. Then one of the Imaun's crew, who had only come in contact with the Icarus, was taken with the disease, which spread through the latter ship, attacking thirty-eight persons, of whom seventeen died. Again, the ship Hydra's boats were for two nights employed in rowing guard round the Icarus, and within two days yellow fever attacked the crew of the Hydra, and was only extinguished by taking the vessel to the northern latitude of Halifax, Nova Scotia. Finally, when the Imaun became the seat of yellow fever, as above stated, a number of supernumeraries were removed from her to the Barracouta, but as much as possible segregated from the crew. Nevertheless, cases of yellow fever soon appeared in this ship, which, like the Hydra, had to run into a more northern latitude. *From one centre the disease radiated in three directions, and at each spot a fresh centre of infection and death was established.* Infection by yellow fever being thus undoubted, it follows that the landing of sick, and their dispersion on shore—or in other words, the abolishment of quarantine—can only tend to multiply circles of epidemic disease.

There are, however, facts which prove that the spread of epidemic disease is not a *sine quâ non* on the landing of sick from yellow fever and similar maladies. It is on such facts that the opponents of quarantine base their arguments against the continuance of the system; but if it be proved, as it certainly is, by Dr. Bryson's report above sketched, that yellow fever has *in any one instance* radiated from a centre, it forms a sufficient reason against the total abolishment of quarantine regulations.

Some of the facts against the system of quarantine on which its opponents base their arguments, are as under, from the ' Report of the Committee of the National Association on Quarantine to the Board of Trade, 1860.'

Showing the evil of the keeping of crews on board ship after disease has broken out, and the benefit of landing them, the Vice-Consul of Galatz states, during the whole time quarantine existed there, about twenty-four or twenty-five years, no case of plague occurred in the lazaretto. But it is on record that the plague was on board a vessel somewhere about 1834, and that all the crew died excepting one man.

Some individuals "see but what they list." In the above extract, I confess I can only discern a reason for the continuance of quarantine. Had that ship's crew been landed, it is more than possible that Galatz would not have remained free from plague for a quarter of a century.

The following extracts are more to the purpose :—" In 1852, Her Majesty's ship Dauntless, with thirty-three cases of yellow fever on board, was admitted at once at Tardeo to Prategne, and removed to St. Anne's Hospital. They all recovered, and no one took it."

Vice-Consul Herbright states (Feb., 1853) :—" The cases of yellow fever at this port (Carthagena) have been exclusively confined to persons landed from the royal mail steamers, and have in no way affected the health of the town."

In 1855 yellow fever raged at Norfolk, and Portsmouth and the country people forbid fugitives. The railways were also stopped ; steamers, however, plied from Norfolk to Baltimore, 180 miles distant, and on the opposite side of the James River. Several hundreds took quarters in Baltimore hotels, but there was no contagion.

It is not, however, stated if any really sick with yellow fever were allowed to make the passage to Baltimore.

Another important fact, apparently afforded by the same report of the National Association, is that quarantine, even when rigidly enforced, has not proved preventive *of other diseases* of the contagious nature, such as smallpox and the exanthemata, as was demonstrated during the epidemic of the former in Malta in 1829 and 1835.

It is, however, an admitted fact, that the exanthemata are less under sanitary control than any other class of zymotic diseases.

The well-remembered case of the Eclair is also quoted as an example of the same kind. Dr. Milroy states, on the arrival of this ship off the English coast, one half of the crew had perished. The surviving medical officer urged the immediate landing of the crew, and Sir J. Richardson, the physician of Haslar Hospital, expressed his readiness to receive them. The ship was, however, ordered to perform a lengthened quarantine, with some fresh volunteers on board, one of whom died, as well as many of the remaining crew.

It is also stated that a noticeable peculiarity is evident, viz., "that the more liberal a government and the freer its institutions, the fewer and less stringent are its quarantine restrictions. In the Baltic states, Sweden, Denmark, Prussia, and Holland, quarantine regulations may be considered a dead letter.

It must, however, be recollected that Sweden, Denmark,

Holland, &c., are situated in a latitude considerably to the north of the locality where the chief diseases against which quarantine regulations are directed have their origin. Indeed, in the instance of yellow fever, *nothing extinguishes the malady so quickly as running into a cold climate.* This has been demonstrated so many times to be a speedy and safe remedy, that I fully concur in the forcible remarks of Dr. McWilliam : " The duty of commanding officers should be defined, and they should be bound to adopt this procedure as the best, and indeed the only means of arresting yellow fever."

Thus it is evident that quarantine regulations are of less importance in northern countries than in those situated nearer the equator; while the absence of plague, yellow fever, and cholera is the exception, where sanitary matters are so frequently neglected, and where therefore the required concomitants exist, for the dissemination and origin of such diseases.

Hence the sick from the Éclair might have been landed with safety. In the climate of England, and in the midst of the sanitary regulations of Haslar, there would have been less danger of the infection spreading than was the case, when last year the typhus fever patients were received into the crowded city of Liverpool from the Egyptian frigate which came there. Owing undoubtedly to strict sanitary regulations, this typhus fever did not become epidemic in the full sense of the term, although it is sufficiently evident that Liverpool is not one of those cities where typhus fever cannot exist.

Had these sick been landed in a tropical port, allowed to pass into all parts of the narrow and crowded thoroughfares of an eastern city, or even located in the apology for hospitals generally found in foreign tropical lands, most medical men acquainted with eastern habits would have little difficulty in foretelling the result.

Thus, although quarantine may be without risk abolished in northern climates, it must ever remain as the best, although confessedly a poor safeguard, of tropical nations. There is, however, no reason why it should not be freed as much as possible from all that is vexatious, costly, or inhuman ; and no candid person who has considered the subject but must feel convinced of the innumerable just causes of complaint which the present system entails.

Sir J. Bowring, so long since as 1841, stated in the House of Commons, that the loss from quarantine in the Mediterranean alone, was not less than two or three million sterling per annum.

The following are two out of many absurd regulations : " All vessels leaving Havannah for Spain between May and October, must proceed to Vigo and there perform a quarantine of fourteen days, *although no yellow fever was at Havannah at the time of departure.*"

At New Orleans all arrivals from Rio Janeiro, the West Indies, and Gulf of Mexico, are liable to quarantine of not less than twelve days, whether the bills of health from those places are clean or foul.

Again, the treatment of passengers is not only vexatious, but positively indecent, and tending to induce disease.

In the 'Times' January 10th, 1854, Mr. Ewart states, that at Naples all the passengers, ladies and gentlemen, were placed in one room, only separated from convicts by a low wall.

Dr. Robertson describes the Beyrout lazaretto as a filthy and damp room, wretched and unhealthy, situated on a low swampy ground.

Dr. Davy relates something worse. At Costanga on the Black Sea, travellers are located in a low hut, not unlike an Irish cabin, without any means of lighting, except the door. This, Dr. Davy remarks, is " in a climate as severe in winter as that described by Ovid in his 'Tristia,' and where strict and vexatious quarantine regulations cannot possibly be necessitated."

Although quarantine can never be safely abolished in

tropical countries, it may be so modified as to change its character altogether, rendering it as little annoying and hurtful as possible, and yet as defensive as circumstances will permit in the way of protection against the spread of infectious disease.

Lastly, if from paramount considerations it ever happens that sick are landed in warm climates where yellow fever for instance can continue to exist, *strict segregation becomes a matter of the first importance.* The same segregation is of equal moment when sick from other diseases, as typhus, are landed in more northerly climates, where also such maladies can be disseminated.

Another important question is, Will goods convey contagion? The general opinion is that they will not; that articles of merchandize are incapable of becoming media of the kind; and founded on the fact, that those whose duty it is to air the goods needing depuration, have never contracted disease.

It is also asked, if yellow fever could be introduced as supposed by fomites, how is it that Liverpool has escaped the disease, where at all seasons cargoes of cotton are arriving from the southern ports of the States, one or other of which is so often the seat of fever?

With respect to the latter question it has already been shown, that yellow fever ceases to exist in a cold latitude; and therefore it is but rational to suppose that fomites cannot induce their characteristic effects in such localities; which is a quite sufficient explanation of the exemption of Liverpool from this pest.

Whether, however, the contagious principles, the fomites of disease, are not susceptible of being conveyed from one place to another in favourable climates is by no means so certain.

As regards the exemption from yellow fever of those who air the goods, allowance must be made for the effect of habit in fortifying the system against contagion. Dr. Watson observes, persons who are much and often exposed to these effluvia, are thereby seasoned in some degree to the noxious atmosphere; "just as drunkards and opium eaters become at length impassive under such a dose of their customary stimulus, as would intoxicate or stupify a novice," and as the inhabitants of malarious countries become proof against malaria.

Moreover, cargo is taken from the shore, and placed in the hold of the ship—a locality quite distinct from that part of the vessel in which the crew live, and hence the "bales" not coming in contact with diseased individuals, or even with diseased atmosphere, may not attract "fomites." Again, as the merchandize is swung out of the ship, and conveyed on shore, it becomes exposed on all sides to the vigorous land or sea breeze, which would tend to dissipate any *materies morbi* clinging to the surface.

No one will deny the truth of Dr. Watson's remarks regarding the exciting cause of typhus fever :—" There are a thousand ways in which contagion may be disseminated. It may lurk in a hackney-coach; you may catch the complaint from your neighbour in an omnibus, or at the theatre, or at church; your linen may be impregnated with the subtle poison in the house of your laundress, or your coat may convey it from the workshop of your tailor. Even if we *doubt* about the contagiousness of the disease, we are bound to act as if we had no doubt on the subject."

It is true that Sir Gilbert Blane did not believe in the contagious nature of fever, if proper arrangements were made for ventilation and cleanliness; and this view of the doctrine of contagion has been also successfully maintained by Dr. Alison, of Edinburgh, and is the one Miss Nightingale adopts at page 93 of her work 'On Hospitals.'

However correct this may *be when sanitary matters are attended to, it is not true when such are neglected ; and* hence merchandise arriving, as it must often do, from localities in which every defect of sanitation exists, exposed as it frequently is to direct contact with the epidemically diseased, cannot be free from the *suspicion* of carrying with it the *materies morbi* of such maladies, and of thus becoming a new centre of infection when the climate, or predisposed constitution of those among whom it may be placed, favours the development of the poison.

THE END.

INDEX.

A.

B.

C.

T.

Z.

ERRATA.

Page 164, line 12, *for* inorganic, *read* organic.

„ 203, „ 6, *for* Dr. Heming, *read* Dr. Fleming.